SOIL AQUIFER TREATMENT: ASSESSMENT AND APPLICABILITY OF PRIMARY EFFLUENT REUSE IN DEVELOPING COUNTRIES

Soil Aquifer Treatment
Assessment and Applicability of Primary Effluent Reuse in Developing Countries

DISSERTATION

Submitted in fulfilment of the requirements of
the Board for Doctorates of Delft University of Technology
and of the Academic Board of the UNESCO-IHE Institute for Water Education
for the Degree of DOCTOR
to be defended in public
on Tuesday, June17, 2014 at 15:00 hours
in Delft, the Netherlands

by

Chol Deng Thon ABEL

Master of Science
UNESCO-IHE Institute for Water Education, Delft, the Netherlands

Born in Malakal, South Sudan

CRC Press
Taylor & Francis Group
Boca Raton London New York

CRC Press is an imprint of the
Taylor & Francis Group, an **informa** business

A BALKEMA BOOK

This dissertation has been approved by the promotor:
Prof. dr. M. D. Kennedy

Composition of the Doctoral Committee:

Chairman:	Rector Magnificus TU Delft
Vice-Chairman:	Rector UNESCO-IHE
Prof. dr. M. D. Kennedy	UNESCO-IHE/Delft University of Technology, promotor
Dr. ir. S. K. Sharma	UNESCO-IHE, copromotor
Prof. dr. ir. J. P. van der Hoek	Delft University of Technology
Prof. dr. ir. T. Wintgens	University of Applied Sciences of Northwestern Switzerland, Muttenz, Switzerland
Prof. dr. S. R. Asolekar	Indian Institute of Technology, Bombay, India
Prof. dr. M. Salgot de Marçay	University of Barcelona, Barcelona, Spain
Prof. dr. ir. L. C. Rietveld	Delft University of Technology (reserve)

First issued in hardback 2018

CRC Press/Balkema is an imprint of the Taylor & Francis Group, an informa business

© 2014, Chol Deng Thon Abel

Published by:
CRC Press/Balkema
PO Box 11320, 2301 EH Leiden, the Netherlands
e-mail: Pub.NL@taylorandfrancis.com
www.crcpress.com – www.taylorandfrancis.co.uk

ISBN 13: 978-1-138-37324-2 (hbk)
ISBN 13: 978-1-138-02673-5 (pbk)

Dedication

To my late father Deng Thon Abel
Hopefully this may be considered as a belated fulfilment of your advice to me to
pursue education

SUMMARY

Unplanned disposal of untreated or inadequately treated wastewater to lakes, streams and land is globally increasing at staggering volumes, especially in developing countries due to rapid population growth, urbanization and lack of investments to construct, operate and maintain conventional wastewater treatment plants (WWTPs). Furthermore, the majority of existing WWPTs (if any) are aging and overloaded since they were designed to serve small fractions of the population that are currently being served. On the other hand, there is increasing water scarcity in different parts of the world and strong competition for water among different sectors. As a consequence, development and implementation of cost effective and environmentally sound treatment technologies with low energy and chemical footprint are desired to alleviate surface water pollution and provide effective integrated water resources management through water reuse. Planned land applications of effluents such as soil aquifer treatment (SAT) have the potential to treat wastewater effluents for subsequent reuse.

SAT is a geo-purification system that utilizes physical, chemical and biological processes during infiltration of wastewater effluent through soil strata to improve water quality. Treatment benefits are initially achieved during vertical infiltration of wastewater effluent through the unsaturated zone and eventually during its horizontal movement in the saturated zone before it is abstracted again from a recovery well. Although SAT has been employed for further treatment and reuse of wastewater effluent in various sites around the world, most experience accumulated is site specific and there are no appropriate tools and methods for knowledge transfer to replicate this experience at new schemes. Furthermore, most SAT sites in developed countries use secondary and tertiary effluents contrary to developing countries in which these levels of treatment are not achieved due to high investment and operation costs. SAT employing primary effluent (PE) is an attractive option for developing countries since wastewater treatment up to this level is cost effective and does not require considerable wastewater treatment plant operator expertise. Nevertheless, little information is available on the use of this type of wastewater effluent for SAT. Therefore, research is needed to understand the fate of suspended solids, organic matter, nutrients, organic micropollutants and pathogens under various process conditions during SAT of PE. Additionally, it is of principle importance to develop a framework and decision support tools through which implementation of new SAT projects could be successfully undertaken.

Laboratory-scale soil columns and batch reactors as well as analysis of data collected from the literature on laboratory experiments, pilot and full-scale SAT sites were performed to establish a clear understanding of SAT performance. The effects of temperature change, redox conditions, soil type, hydraulic loading rate (HLR), pre-treatment of PE, biological activity, wetting and drying cycles on removal of selected contaminants from PE were investigated.

Laboratory-scale soil columns were used to investigate the effect of PE pre-treatment (prior to application to SAT) on the removal of suspended solids (SS), bulk organic matter (measured as dissolved organic carbon (DOC)), nutrients (nitrogen and phosphorus) and pathogen indicators. Two coagulants namely, aluminium sulfate and iron chloride were tested. Experimental results revealed no difference in the overall removal of SS (which levelled off at ~90%) during infiltration of coagulated and non-coagulated PE. However, coagulation-infiltration increased the removal of bulk organic matter, phosphorus and pathogens indicators respectively from 16 to ~70%, 80 to >98% and 2.6 to >4 $\log_{10}$ units (with low removals achieved during infiltration only). Both coagulants could be equally employed to improve the overall performance of SAT system and reduce area requirements.

Effect of soil type and HLR on removal of bulk organic matter and nitrogen was explored using a 5 m long soil columns packed with silica sand and dune sand. No significant difference in DOC removal (~46%) was observed when the HLR was reduced from 1.25 to 0.625 m/d. However, removal of ammonium-nitrogen was 50% higher at HLR of 0.625 mg/d compared to HLR of 0.625 mg/L. Furthermore, ammonium-nitrogen removal in dune sand column was 10% higher than the removal in silica sand column. In conclusion, SAT system with relatively fine soil particles and operated at relatively low HLR provides better removal of ammonium-nitrogen. Nevertheless, such system requires much frequent drying and scraping of soil surface.

Removal of SS, bulk organic matter, nitrogen and pathogens indicators from PE was investigated at continuous wetting and varying wetting/drying periods using a 4.2 m long soil column. No significant increase in the removal of SS (~90%) and DOC (50-60%) was observed with increase in the drying period. Nevertheless, notable increase in removal of ammonium-nitrogen and pathogens indicators was observed with increase in drying period. Ammonium-nitrogen removal increased from as low as 20% at continuous wetting to 98% at drying period of 6.4 days whereas removal of *E. coli* and *total coliforms* increased from 2.5 $\log_{10}$ units under continuous wetting to >4 $\log_{10}$ units at 6.4 days drying period. In summary, removal of SS and DOC was independent of the length of the drying period whereas removal of nitrogen, *E. Coli* and *total coliforms* increased progressively as the length of the drying period increased.

The influence of biological activity on the removal of bulk organic matter, nitrogen and selected pharmaceutically active compounds (PhACs) from PE was studied in laboratory-scale batch reactors. Biological activity (measured as adenosine triphosphate (ATP) correlated positively with removal of DOC, which increased progressively from 14% in blank reactors to 75% in the reactor with the highest biological activity. Likewise, removal of ammonium-nitrogen increased with biological activity from 10 to 95%. While removal of hydrophilic neutral compounds (octanol-water partition coefficient log K_{ow} <2) of phenacetin, paracetamol and caffeine was independent of the extent of biological activity and was >90%, removal of pentoxifylline was dependent on biological activity and length of reactor ripening period. On the other hand, removal of gemfibrozil, diclofenac and bezafibrate increased from less than 10% in blank and control reactors to >80% in biological active reactors implying dependence on biological activity. Removal of clofibric acid and carbamazepine was <50% in most reactors suggesting that removal of these compounds was not dependent on biological activity. Conclusionally, removal of DOC correlated positively with the extent of biological activity. Likwise, removal of PhACs gemfibrozil, diclofenac, bezafibrate, ibuprofen, naproxen and ketoprofen increased notably with biological activity, whereas carbamazepine and clofibric acid were found persistent irrespective of the extent of the biological activity in the reactor.

The effect of temperature and redox conditions on the removal of bulk organic matter, nitrogen, phosphorus and pathogen indicators was examined using laboratory-scale soil columns and batch reactors. While an average DOC removal of 17% was achieved in soil columns at 5°C, the removal increased by 10% for every 5°C increase in temperature over the range of 15-25°C, with DOC removal of 69% achieved at 25°C. Furthermore, aerobic soil columns exhibited a DOC removal 15% higher than that in anoxic columns, while aerobic batch reactors showed DOC removal 8% higher than the corresponding anoxic batch experiments. Ammonium-nitrogen removal >99% was observed at 20°C and 25°C, whereas the removal decreased substantially to 9% at 5°C. While ammonium-nitrogen was attenuated by 99% in aerobic batch reactors at room temperature, anoxic experiments under similar conditions resulted in 12% ammonium-nitrogen reduction. In light of these findings, SAT system operated at high temperature in summer will provide high removal rates of DOC, nitrogen, *E. Coli* and *total coliforms* from PE compared to low winter temperature. Inadequate aeration of SAT system due to short drying periods could result in poor reduction of ammonium-nitrogen.

Current SAT sites under operation around the globe tend to focus on the operational aspects to meet the reuse quality requirements. As a result, development of assessment tools that help implement SAT technology at new sites based on the experience gained at the sites currently under operation has not been addressed. In this study, a framework and tools for SAT implementation were developed for different users ranging from decision makers, planners, engineers and SAT site operators. SAT pre-feasibility tool covered institutional, legal, socio-political and technical requirements while site identification, design, operation and maintenance tools were developed. Furthermore, a water quality prediction model was developed to estimate the potential removal of DOC, nitrogen, phosphorus, bacteria and viruses based on wastewater effluent characteristics, pre-treatment and travel distance. The model is especially useful to assess the need for post-treatment in order to meet reclaimed water quality requirements for reuse and assists in estimation of the total investment cost required to incorporate any post-treatment.

This thesis investigated the potential of using SAT technology for further treatment and reuse of PE through experimental work and the development of assessment tools suitable for different stages of SAT coupled with a water quality prediction model. Although the tools and water quality prediction model were developed, tested and validated using data from laboratory experiments, pilot and SAT sites situated in developing countries, these tools and model are generic and could be easily adapted to suit different sites in developing countries. The thesis provides a comprehensive methodology that will be useful for decision makers, planners and engineers to develop and operate new SAT schemes especially in developing countries where SAT (using PE) has not been exploited to its maximum potential.

ACKNOWLEDGEMENTS

First and foremost, I am grateful to UNESCO-IHE Partnership Research Fund (UPaRF) that facilitated this study under the financial support of NATSYS project.

I would like to express my gratitude to my promoter Prof. Dr. Maria D. Kennedy for sharing her vast expert and scientific knowledge with me, not only on topics related to my research, but also sharing good career advice. I gratefully acknowledge with thanks, the meticulous supervision and guidance provided by my supervisor Assoc. Prof. Dr. Saroj K. Sharma. This study would not have come to completion without your conscientious support and insightful comments. I also owe a special thanks to Prof. Dr. Gary L. Amy for his invaluable comments and timely feedback over the course of the study.

My sincere thanks and appreciation go to the staff of the UNESCO-IHE Environmental Engineering Laboratory, namely: Fred Kruis, Frank Wiegman, Peter Heerings, Lyzette Robbemont, Ferdi Battes, Berend Lolkema and Don van Galen for their friendly support with analysis in the laboratory. I would also like to extend my gratitude to Jolanda Boots (Ph.D. Fellowship and Admission Officer), Ewout Heeringa, Ed van der Hoop and Eric Pluim. I sincerely acknowledge the help provided by Berthold Verkleij and Dennis Thijm during collection of wastewater samples. I am grateful to Tanny van der Klis, Chantal Groenendijk, Anique Karsten and Peter Stroo for their impeccable communication work. Special thanks to Dr. Sung Kyu Maeng, Dr. Abraham Mehari, Dr. Yasir Mohammed, Dr. Jan Willem Foppen, Dr. Henk Lubberding, Guy Beaujot, Dr. Kebreab Ghebremichael and Dr. Mariska Ronteltap for their insightful comments and words of encouragement. Dr. Ronteltap, thank you for translating both summary and propositions to Dutch language. I am also very grateful to Prof. Dr. Piet Lens who gave me two books (related to my PhD topic) from his personal book collection.

Throughout the research time, I made some good friends who were very supportive in various ways contributing directly and indirectly to this thesis. I am grateful to Ervin Buçpapaj, Yona Malolo, Joseph Ntelya, Selamwit Mersha and Khalid Al Kubati for their contribution with the experimental work during their MSc study at UNESCO-IHE. I would also want to thank Roman Vortisch, an MSc student from Dresden University of Technology, Germany who conducted part of his MSc research under my supervision for six months at UNESCO-IHE. I gratefully acknowledge with thanks, my Ph.D. colleagues Yasir Ali, Khalid Hassaballah, Micah Mukolwe, Nirajan Dhakal, Sergio Salinas (Dr.), Laurens Welles, Loreen Villacorte (Dr.), Saeed Baghoth (Dr.), Abdulai Salifu, Assiyeh Tabatabai (Dr.), Mulele Nabuyanda, Omar Munyaneza (Dr.), Peter Mawioo, George Lutterodt (Dr.), Mohanasundar Radhakrishnan, Harrison Mutikanga (Dr.), Rohan Jain, Silas Mvulirwenande, Jeremiah Kiptala, Valentine Uwamariya (Dr.), Hans Komakech (Dr.), Frank Massesse, Girma Ebrahim (Dr.) and Jae Chung who made me forget the loneliness of being far from my family.

I owe a debt of gratitude to the South Sudanese community in the Netherlands for their warm hospitality. My sincere thanks and appreciation go particularly to Peter Makoi, Ater Makurthou, Ayuel Kacgor, Maker Makurthou, Rose Abang Kuot, Akuac Ajang, Michael Amol, Deng Barac, Emmanuel Scopus, Ter Bishok, Bella Kodi and Brian Oburak. A special note of thanks goes to the Sudanese community in Delft.

This thesis would not have come into existence had it not been for the support and love granted to me by my family: my beloved mother Nyandueny Deng Awool Kiir, my charming wife Dr. Angeth (Helen) Abraham Adual, my late brother Abiel Deng Thon Abel, my sister Mary Deng Thon Abel, my sister Abuk Deng Thon Abel and my younger brother Aban Deng Thon Abel. Special thanks go to my relatives and extended family members, your contributions during my Ph.D. study have helped me to accomplish it. I dearly and sincerely appreciate your kind help.

Above all, my heartfelt thanks go to God almighty for his protection and providence.

Chol Deng Thon Abel

June 17, 2014

Delft, the Netherlands

TABLE OF CONTENTS

CHAPTER 3   EFFECT OF PRE-TREATMENT OF PRIMARY EFFLUENT USING ALUMINUM SULFATE AND IRON CHLORIDE ON REMOVAL OF SUSPENDED SOLIDS, BULK ORGANIC MATTER, NUTRIENTS AND PATHOGENS INDICATORS   49

CHAPTER 4   IMPACT OF HYDRAULIC LOADING RATE AND SOIL TYPE ON REMOVAL OF BULK ORGANIC MATTER AND NITROGEN FROM PRIMARY EFFLUENT IN LABORATORY-SCALE SOIL AQUIFER TREATMENT SYSTEM   67

LIST OF ACRONYMS

AMB	Active Microbial Biomass
ATP	Adenosine triphosphate
CG	Coagulation
DN	Disinfection
DO	Dissolved Oxygen
DOC	Dissolved Organic Carbon
EBCT	Empty Bed Contact Time
FEEM	Fluorescence Excitation Emission Matrix
FI	Frequency Index
GAC	Granular Activated Carbon
HLR	Hydraulic Loading Rate
HRT	Hydraulic Residence Time
MAR	Managed Aquifer Recharge
MF	Micro-filtration
OCC	Optimum Coagulant Concentration
OMPs	Organic Micropollutants
ORP	Oxidation Reduction Potential
PE	Primary Effluent
PhACs	Pharmaceutically Active Compounds
RSF	Rapid Sand Filtration
SAT	Soil Aquifer Treatment
SE	Secondary Effluent
SP	Settling Pond
SPSS	Statistical Package for the Social Sciences
SS	Suspended Solids
SUVA	Specific Ultra-violet Absorbance
TE	Tertiary Effluent
TOC	Total Dissolved Carbon
UF	Ultra-filtration
WWTP	Wastewater Treatment Plant

LIST OF FIGURES

LIST OF TABLES

CHAPTER 1

INTRODUCTION

SUMMARY

Globally, rapid population growth and urbanization are increasing domestic, agricultural and industrial water demands and diminishing available water resources. On the other hand, the volume of wastewater generated is envisaged to increase in the future and exert stress on existing wastewater treatment facilities. In developed countries, availability of investment, operation and maintenance costs enables these countries to cope with this challenge by expanding their current facilities. Nevertheless, developing countries lack financial resources and technical expertise to develop wastewater treatment facilities to treat wastewater to secondary or tertiary effluent levels and wastewater is either discharged to receiving water bodies after undergoing partial treatment or not treated at all. This suggests that water reuse in developing countries is not only desirable, but imminently inevitable to alleviate adverse health impacts and degradation of receiving water bodies. Pre-treatment of wastewater to the level of primary effluent can be coupled with cost-effective and environmentally sound technology (i.e. soil aquifer treatment) to efficiently reduce pressure on freshwater resources. This thesis explores different aspects of a soil-based natural treatment system for treatment of primary effluent aiming at subsequent reuse.

1.1   BACKGROUND

Water scarcity is considered as one of the challenges faced by human society across the globe (Bdour et al., 2009). The supply of freshwater is limited and cannot meet the growing demand (Wild et al., 2007). Factors like contamination of surface water and groundwater, uneven distribution of water resources, and frequent droughts caused by extreme global weather patterns have severely influenced water scarcity (Asano and Cotruvo, 2004). According to WHO and UNICEF (2013) by the end of 2011, 780 million people in the world did not have access to improved water supply and 2.5 billion people did not have access to improved sanitation. On the other hand, rapid population growth and extension of irrigated agriculture are posing stress on the available water resources. The world population is envisaged to increase from 7.2 billion in 2014 to 9.3 billion in 2050 (UN, 2007), while urban population is projected to increase from 3.4 billion to 6.4 billion in the same period (Corcoran, 2010; UNPD., 2007). This growth in population and urbanization rates will exert more stress on available water resources due to increase in water demand for food production (Corcoran, 2010). Urbanization and industrial expansion may exert severe anthropogenic environmental impact on surface water leading to contamination with a wide range of trace organic compounds (Schmidt et al., 2007). Furthermore, excessive exploitation patterns and pumping rates from groundwater, in excess of natural replenishment, leads to rapid decline in groundwater levels and eventual depletion of groundwater resources (Abel et al., 2013; Asano and Cotruvo, 2004). Wastewater volumes and the need for collection and treatment will proportionally increase with the anticipated growth in urban water supply coverage in these cities since wastewater represents 75%-85% of water supply (Scott et al., 2004). To cope with such problems, an urgent means of "artificial" water storage with suitable facilities is needed (Díaz-Cruz and Barceló, 2008). Wastewater provides a source of water that could extensively reduce exploitation of valuable natural water resource (Drewes and Khan, 2010; Toze, 1997) for non-potable reuse purposes.

In their quest for alternative water sources, several communities in arid and semi-arid regions of the world have considered treated municipal wastewater as an integral part of their water supply options (Quanrud et al., 1996; Asano and Cotruvo, 2004; Guizani et al., 2011). Wastewater in some water scarce southern Africa and Middle East states has become a valuable resource that is used for agriculture, groundwater recharge and urban applications after a polishing treatment phase (Bdour et al., 2009). According to Asano (2007) water reuse serves as a complementary water source which is accessible throughout the year in urban areas for various reuse purposes. Water reuse is frequently practiced as a method for water resources management (Guizani et al., 2011; Vigneswaran and Sundaravadivel, 2004). It has many benefits such as protection of water resources, prevention of coastal pollution, recovery of nutrients for agriculture, augmentation of river water flow, saving in wastewater treatment and groundwater recharge (Angelakis and Bontoux, 2001; Huertas et al., 2008). Water reuse applications encompass; agricultural irrigation, landscape irrigation, groundwater recharge, industrial reuse, environmental and recreational uses non-potable urban uses and indirect or direct potable reuse (Asano,

2002; Huertas et al., 2008). Quanrud et al. (2003) asserted that several countries have acknowledged indirect potable reuse systems that percolate reclaimed water and retard its reuse by undergoing aquifer storage. In Belgium, Mexico, United States of America (USA) and Singapore, planned indirect potable reuse is employed through treatment of wastewater effluents to augment drinking water supplies (Drewes and Khan, 2010). On the other hand, Windhoek, Namibia, has been practising direct potable reuse of highly treated wastewater effluent for drinking water supply since 1969 (du Pisani, 2006; Le-Minh et al., 2010). Nevertheless, the largest indirect reuse project for non-potable purpose is in Shafdan, Israel where 65-75% of the generated wastewater is reclaimed through land application and used predominantly for irrigation agriculutre (Nadav et al., 2012).

Land has been long used for treatment and disposal of wastewater (Duan et al., 2010; McDowell-Boyer et al., 1986). Land treatment through wastewater spreading to the soil dates back to as early as 2600 BC during Minoan Civilization (Angelakis and Spyridakis, 1996). Wastewater land application is not only employed for municipal wastewater treatment and disposal, but also provides a wide spectrum of environmental, economic and social benefits (Duan and Fedler, 2007; Duan et al., 2010). It is a threefold application that serves: (1) providing reliable treatment of wastewater to meet water quality requirements for intended reuse, (2) protecting public health and (3) obtaining acceptance (Asano, 2002) of the population served. However, chemical, geological, geochemical, and public health parameters in conjunction with land-use ecology should be intensively studied before potential reuse of treated wastewater to ensure safe water reuse (Kalavrouziotis and Apostolopoulos, 2007). The quality of wastewater effluent infiltrated during land based water treatment is improved through filtration, adsorption, chemical and biodegradation processes in the aerated unsaturated zone and dispersion and dilution in the underlying aquifer (Nema et al., 2001). The land treatment in which both soil and aquifer participate in wastewater renovation is called soil aquifer treatment (SAT). SAT, riverbank filtration (RBF) and artificial recharge and recovery (ARR) known collectively (among others) as managed aquifer recharge (MAR), are natural processes used in drinking water augmentation projects that could produce potable water from water sources under influence of wastewater (Rauch-Williams et al., 2010). These wastewater land application systems reduce the pressure on freshwater resources in arid and semi-arid areas (Heidarpour et al., 2007). MAR is a planned recharge of water to aquifers for later recovery or for environmental advantages (Dillon et al., 2010).

1.2   THE NEED FOR RESEARCH

Water represents 99.9% of the total volume of municipal wastewater, while suspended and dissolved organic and inorganic solids represent a very small portion (Pescod, 1992). High water content in wastewater makes its collection, treatment and reuse, a viable option to introduce integrated urban water management and provide a reliable new water source. Nevertheless, many developing countries lack adequate wastewater treatment facilities, reliable power supply and skilled personnel to run and maintain these facilities (Horan, 1990). High investment, operational and maintenance costs of conventional wastewater treatment technologies make construction of these facilities an expensive option that does not suit developing countries (Hussain et al., 2007; Westerhoff and Pinney, 2000). Therefore, wastewater in many of these countries is either not properly treated or not treated at all before it is discharged back to the water cycle (Wild et al., 2007). In India, only 24% of domestic and industrial wastewater is treated while only 2% of the same wastewater type is treated in Pakistan (Mexico, 2003). Additionally, West African cities wastewater treatment facilities receive and treat less than 10% of the generated wastewater through sewerage system (Drechsel et al., 2006) whereas Latin America treats only 7% of its wastewater. As a consequence, considerable volumes of untreated wastewater effluent are channelled back to receiving water bodies leading to water quality deterioration. Untreated municipal and industrial effluent poses a serious threat to population health in some of these regions (Wild et al., 2007). Such detrimental effects could be reduced by using a cost-effective and environmentally friendly technology with low energy requirements to polish these effluents for reuse applications. Treatment benefits can be maximized if wastewater is pre-treated to the level of primary effluent before undergoing further treatment. SAT is among the technologies that can reliably and consistently produce treated wastewater of acceptable quality.

SAT provides additional treatment to primary, secondary and tertiary effluents from wastewater treatment plants (WWTPs) for reuse purposes (Crites et al., 2006; Fox et al., 2001; Nema et al., 2001; Sharma et al., 2011; Wilson et al., 1995). As a result of poor wastewater quality due to inadequate treatment in vast majority of developing countries, application of primary effluent (PE) in SAT systems in these countries has the potential to augment existing water resources to meet the growing water demand and enhance water availability for different uses (Sharma et al., 2011). PE is the partially treated wastewater after removal of floating materials, grit, settleable organic and inorganic solids through screening, skimming and sedimentation (Haruvy, 1997; Pescod, 1992). The use of PE in SAT also provides an economic benefit since wastewater treatment to PE level does not require sizable investment compared to secondary effluent and tertiary levels and SAT does not require extensive use of energy and chemicals. However, PE is characterized with high ammonium, high sediment load, low nitrate and relatively high phosphorus concentrations (Abel et a., 2012; Ho et al., 1992). Besides, organic carbon is a major water quality concern in SAT schemes that involve indirect potable reuse of the reclaimed water (Drewes et al., 2006).

SAT has been used for the treatment of PE, but the effect of water quality parameters and climatic conditions is not fully known. Current SAT experiences in developed countries are site specific and lack appropriate tools to facilitate knowledge transfer to new schemes in the developing world are lacking. No information is available on the fate of organic micro-pollutants (OMPs) in PE during SAT. Furthermore, effect of temperature, redox conditions, soil properties, PE pre-treatment as well as hydraulic loading conditions (wetting and drying) on performance of SAT using PE is not well documented. Therefore, this research seeks to bridge the knowledge gap on the use of PE in SAT by using real PE from different

WWTPs in laboratory experimental setups (soil columns and batch reactors) simulating SAT to probe efficiency of SAT in removal of multi contaminants and explore its applicability and suitability for the treatment of PE in developing countries. The research addresses the effects of coagulation, hydraulic loading rate (HLR), soil type, temperature, redox conditions and physical and biological mechanisms on the removal of suspended solids, bulk organic matter, nitrogen, phosphorus, organic micro-pollutants and pathogenic indicators in PE. Furthermore, the research will improve current knowledge on site selection, screening, design, operation and maintenance by developing a framework necessary to facilitate this process in new SAT schemes. As part of this research work, an excel-based modelling tool was developed using the available data on SAT to enable planner and SAT proponents to predict the removal of potential contaminants of interest at SAT sites from different wastewater effluents and soil types based on the distance between recovery wells and infiltration basin.

1.3   RESEARCH OBJECTIVES

The overarching research goal was to investigate the suitability and viability of SAT for treatment of PE under various climatic and processes conditions for reuse.

In order to achieve the above overall goal, some specific objectives have been identified as follows:

- To assess the influence of PE pre-treatment and infiltration on removal efficiency of SAT for suspended solids, bulk organic matter, nutrients and pathogens indicators.
- To investigate the impact of soil type and hydraulic loading rate on attenuation of bulk organic matter and nitrogen from PE during soil passage.
- To probe the effects of operating process conditions (wetting/drying) on attenuation of suspended solids, bulk organic matter, nutrients (nitrogen and phosphorus) and pathogens indicators in a SAT system.
- To investigate the impact of physical and biological removal mechanisms on attenuation of bulk organic matter, nitrogen and pharmaceutically active compounds (PhACs) in MAR.
- To explore the influence of temperature variation and redox conditions on removal of bulk organic matter, nutrients and pathogens from PE in a SAT system.
- To develop framework and tools for site selection, design, operation and maintenance for SAT systems.
- To develop a water quality prediction model that is used to estimate removal of contaminants in a SAT system.

1.4 OUTLINE OF THE THESIS

This thesis is organized in nine chapters, each addressing one or more of the research objectives. A short description of each chapter is presented below.

Chapter 1 introduces the challenges encountered to reliably supply freshwater and provide adequate sanitation due to rapid population growth.

A comprehensive review of SAT technology presents system definition, types, operation and maintenance in Chapter 2.

Chapter 3 deals with effects of pre-treatment of PE using aluminium sulfate and iron chloride coagulants on SAT performance.

Chapter 4 analyzes the effects of hydraulic loading rate and soil type on efficiency of SAT to remove bulk organic matter and nitrogen.

Chapter 5 looks at the influence of various operating conditions (HLR, wetting and drying periods) on attenuation of suspended solids, bulk organic matter, nitrogen and pathogens indicators during soil passage.

In Chapter 6, the effect of biological activity on removal of bulk organic matter, nitrogen and pharmaceutically active compounds (PhACs) is presented.

Chapter 7 probes the impact of temperature variation and redox conditions on reduction of bulk organic matter, nutrients (nitrogen and phosphorus) and pathogens indicators in SAT.

Chapter 8 sets out a framework and tools for SAT that are oriented towards helping planners, engineers and operators to select, design and operate new SAT schemes. Furthermore, it provides a water quality model that predicts SAT water quality based on the type of wastewater effluent, pre-treatment and travel distance.

Chapter 9 outlines the thesis summary, conclusions and prospects for further research. This last chapter draws together all the findings from different chapters of this thesis and set forth future prospects for more research on SAT.

1.5 REFERENCES

Abel, C. D. T., Sharma, S. K., Buçpapaj, E. and Kennedy, M. D. (2013). Impact of hydraulic loading rate and media type on removal of bulk organic matter and nitrogen from primary ef uent in a laboratory-scale soil aquifer treatment. *Water Science and Technology,* **68**(1), 217-226.

Abel, C. D. T., Sharma, S. K., Malolo, Y. N., Maeng, S. K., Kennedy, M. D. and Amy, G. L. (2012). Attenuation of Bulk Organic Matter, Nutrients (N and P), and Pathogen Indicators During Soil Passage: Effect of Temperature and Redox Conditions in Simulated Soil Aquifer Treatment (SAT). *Water, Air and Soil Pollution,* **223**, 5205-5220.

Angelakis, A. and Bontoux, L. (2001). Wastewater reclamation and reuse in Eureau countries. *Water Policy,* **3**(1), 47-59.

Angelakis, A. N. and Spyridakis, S. V. (1996). The status of water resources in Minoan times: A preliminary study. *NATO ASI Series I Global Environmental Change,* **36**, 161-192.

Asano, T. (2002). Water from (waste) water- the dependable water resource. *Water Science and Technology,* **45**(8), 24.

Asano, T. (2007). *Water Reuse: Issues, Technologies, and Applications.* McGraw-Hill Professional.

Asano, T. and Cotruvo, J. (2004). Groundwater recharge with reclaimed municipal wastewater: health and regulatory considerations. *Water Research,* **38**(8), 1941-1951.

Bdour, A., Hamdi, M. and Tarawneh, Z. (2009). Perspectives on sustainable wastewater treatment technologies and reuse options in the urban areas of the Mediterranean region. *Desalination,* **237**(1-3), 162-174.

Corcoran, E. (2010). *Sick Water?: The Central Role of Wastewater Management in Sustainable Development: a Rapid Response Assessment.* UNEP/Earthprint.

Crites, R. W., Reed, S. C. and Middlebrooks, E. J. (2006). *Natural Wastewater Treatment Systems.* CRC Press, Boca Raton, Florida, USA, pp 413-426.

Díaz-Cruz, M. and Barceló, D. (2008). Trace organic chemicals contamination in ground water recharge. *Chemosphere,* **72**(3), 333-342.

Dillon, P., Toze, S., Page, D., Vanderzalm, J., Bekele, E., Sidhu, J. and Rinck-Pfeiffer, S. (2010). Managed aquifer recharge: rediscovering nature as a leading edge technology. *Water Science and Technology,* **62**(10), 2338-2345.

Drechsel, P., Graefe, S., Sonou, M. and Cofie, O. O. (2006). Informal Irrigation in Urban West Africa: An Overview. In: Research Report. 102, IWMI.

Drewes, J. and Khan, S. (2010). Water reuse for drinking water augmentation. In: Edzwald, J. (Ed.), Water Quality and Treatment Handbook, Sixth ed. McGraw-Hill.

Drewes, J., Quanrud, D., Amy, G. and Westerhoff, P. (2006). Character of organic matter in soil-aquifer treatment systems. *Journal of Environmental Engineering,* **132**, 1447-1458.

du Pisani, P. L. (2006). Direct reclamation of potable water at Windhoek's Goreangab reclamation plant. *Desalination,* **188**(1), 79-88.

Duan, R. and Fedler, C. (2007). Quality and Quantity of Leachate in Land Application Systems. *ASABE Annual International Meeting, Minneapolis Convention Center, Minneapolis, MN, June17-20, 2007.* Paper number: 074079.

Duan, R., Sheppard, C. D. and Fedler, C. B. (2010). Short-term effects of wastewater land application on soil chemical properties. *Water, Air and Soil Pollution,* **211**(1), 165-176.

Fox, P., Houston, S., Westerhoff, P., Drewes, J., Nellor, M., Yanko, B., Baird, R., Rincon, M., Arnold, R. and Lansey, K. (2001). An Investigation of Soil Aquifer

Treatment for Sustainable Water Reuse. *Research Project Summary of the National Center for Sustainable Water Supply (NCSWS)*, Tempe, Arizona, USA.

Guizani M, Kato H, Funamizu N (2011). Assessing the removal potential of soil-aquifer treatment system (soil column) for endotoxin. *Journal of Environmental Monitoring* 13(6):1716-1722.

Haruvy, N. (1997). Agricultural reuse of wastewater: nation-wide cost-benefit analysis. *Agriculture, Ecosystems and Environment*, **66**(2), 113-119.

Heidarpour, M., Mostafazadeh-Fard, B., Abedi Koupai, J. and Malekian, R. (2007). The effects of treated wastewater on soil chemical properties using subsurface and surface irrigation methods. *Agricultural Water Management*, **90**(1), 87-94.

Ho, G., Gibbs, R., Mathew, K. and Parker, W. (1992). Groundwater recharge of sewage effluent through amended sand. *Water Research*, **26**(3), 285-293.

Horan, N. (1990). *Biological Wastewater Treatment Systems: Theory and Operation*. John Wiley and Sons, New York, USA, pp. 310.

Huertas, E., Salgot, M., Hollender, J., Weber, S., Dott, W., Khan, S., Schäfer, A., Messalem, R., Bis, B. and Aharoni, A. (2008). Key objectives for water reuse concepts. *Desalination*, **218**(1-3), 120-131.

Hussain, S., Aziz, H., Isa, M. and Adlan, M. (2007). Physico-chemical method for ammonia removal from synthetic wastewater using limestone and GAC in batch and column studies. *Bioresource Technology*, **98**(4), 874-880.

Kalavrouziotis, I. and Apostolopoulos, C. (2007). An integrated environmental plan for the reuse of treated wastewater effluents from WWTP in urban areas. *Building and Environment*, **42**(4), 1862-1868.

Le-Minh, N., Khan, S. J., Drewes, J. E. and Stuetz, R. M. (2010). Fate of antibiotics during municipal water recycling treatment processes. *Water Research*, **44**(15), 4295-4323.

McDowell-Boyer, L., Hunt, J. and Sitar, N. (1986). Particle transport through porous media. *Water Resources Research*, **22**(13), 1901-1921.

Mexico, G. (2003). *Confronting the Realities of Wastewater Use in Agriculture*, IWMI, Colombo, Sri Lanka.

Nadav, I., Arye, G., Tarchitzky, J. and Chen, Y. (2012). Enhanced infiltration regime for treated-wastewater purification in soil aquifer treatment (SAT). *Journal of Hydrology*, **420–421**, 275-283.

Nema, P., Ojha, C., Kumar, A. and Khanna, P. (2001). Techno-economic evaluation of soil-aquifer treatment using primary effluent at Ahmedabad, India. *Water Research*, **35**(9), 2179-2190.

Pescod, M. (1992). Wastewater Treatment and Use in Agriculture, Food and Agriculture Organization Irrigation and Drainage Paper 47. Rome

Quanrud, D., Arnold, R., Wilson, L. and Conklin, M. (1996). Effect of soil type on water quality improvement during soil aquifer treatment. *Water Science and Technology*, **33**(10), 419-432.

Quanrud, D., Hafer, J., Karpiscak, M., Zhang, J., Lansey, K. and Arnold, R. (2003). Fate of organics during soil-aquifer treatment: sustainability of removals in the field. *Water Research*, **37**(14), 3401-3411.

Rauch-Williams, T., Hoppe-Jones, C. and Drewes, J. (2010). The role of organic matter in the removal of emerging trace organic chemicals during managed aquifer recharge. *Water Research,* **44**(2), 449-460.

Schmidt, C., Lange, F. and Brauch, H. (2007). Characteristics and evaluation of natural attenuation processes for organic micropollutant removal during riverbank filtration. *Water Science and Technology: Water Supply,* **7**(3), 1-7.

Scott, C., Faruqui, N. and Raschid-Sally, L. (2004). *Wastewater Use in Irrigated Agriculture: Coordinating the Livelihood and Environmental Realities.* CAB International Publishing, Wallingford, UK.

Sharma, S. K., Hussen, M. and Amy, G. L. (2011). Soil aquifer treatment using advanced primary effluent. *Water Science and Technology,* **64**(3), 640-646.

Toze, S. (1997). Microbial Pathogens in Wastewater. CSIRO Land and Water Technical Report, **1**, 97.

UN. (2007). United Nations Water Statistics and pictures https://www.unwater.org/statistics.html (Accessed September 2013), Vol. 2013.

UNPD. (2007). World Urbanization Prospects: The 2007 Revision Population Database, United Nations Population Division, Department of Economic and Social Affairs Brussels, Belgium.

Vigneswaran, S. and Sundaravadivel, M. (2004). Recycle and Reuse of Domestic Wastewater. *Saravanamuthu (Vigi) Vigneswaran], in Encyclopedia of Life Support Systems (EOLSS), Developed under the Auspices of the UNESCO, Eolss Publishers, Oxford UK,[http://www. eolss. net][Retrieved April 24, 2006].*

Westerhoff, P. and Pinney, M. (2000). Dissolved organic carbon transformations during laboratory-scale groundwater recharge using lagoon-treated wastewater. *Waste Management,* **20**(1), 75-83.

WHO and UNICEF. (2013). Progress on Sanitation and Drinking Water: 2013 update. World Health Organization and United Nations Children's Emergency Fund. 9241505397.

Wild, D., Buffle, M.-O. and Hafner-Cai, J. (2007). Water: a market of the future. SAM, SERI. http://www.sam-group.com/2010_water_study_e_tcm794-263789.pdf. Accessed on July 16, 2011..

Wilson, L., Amy, G., Gerba, C., Gordon, H., Johnson, B. and Miller, J. (1995). Water quality changes during soil aquifer treatment of tertiary effluent. *Water Environment Research,* **67**(3), 371-376.

CHAPTER 2

SOIL AQUIFER TREATMENT (SAT): SITE DESIGN, SELECTION, OPERATION AND MAINTENANCE

SUMMARY

Soil aquifer treatment (SAT) is a land-based managed aquifer recharge (MAR) technology which is increasingly adopted as a useful ancillary means to reliably enhance water resources and reduce indiscriminate discharge of treated wastewater to water bodies. During SAT, physical, chemical and biological processes improve the quality of wastewater effluent during its infiltration through soil strata and yield water of acceptable quality for reuse purposes. In order to design and develop a new SAT scheme, numerous factors are considered. Of principle importance during the pre-design phase are the intended use of the water abstracted from SAT recovery wells, public health, economic aspects, regulations and guidelines, socio-political and institutional aspects. However, the design phase focuses on land area availability, site hydrogeology, type of SAT, wastewater effluent pre-treatment, site selection, soil clogging, groundwater mounding, infiltration system design and post-treatment requirements. As part of SAT feasibility study, site investigation is carried out to understand site geological profile and ensure absence of impermeable layers in the unsaturated zone. Furthermore, field and laboratory tests are conducted on potential SAT site to assess groundwater quality, explore soil type, particle size distribution, presence of trace metals and actual infiltration rates.

2.1 INTRODUCTION

SAT is a natural treatment technology which can yield effluent water of adequate quality for indirect water reuse purposes when coupled with other appropriate wastewater treatment technologies (Sharma et al., 2007). It is a geo-purification system in which the aquifer is recharged with partially treated wastewater through unsaturated soil strata before it mixes with the native groundwater (Bdour et al., 2009). Several SAT processes improve water quality during percolation through the unsaturated (vadose) zone (Quanrud et al., 2003) before it got dispersed and diluted in the aquifer (Nema et al., 2001). However, most water quality improvements are obtained during percolation through the vadose zone (Quanrud et al., 1996). Physical, chemical and biological processes retard the water or transform the dissolved contaminants during soil passage leading to mitigation of groundwater pollution (Martin and Koerner, 1984). Organic compounds, nitrogen, phosphorus, suspended solids (SS), trace metals, bacteria and viruses can be effectively removed through sorption, chemical reaction, biotransformation, die-off and predation processes during SAT (Kanarek and Michail, 1996; Zhang et al., 2007).

Though the quality of renovated wastewater through SAT technology is by far better than the influent wastewater, its quality could be slightly different from the native groundwater (Bdour et al., 2009). As such, intrusion of wastewater effluent applied to SAT into groundwater should be avoided by using a small portion of aquifer for SAT (Asano and Cotruvo, 2004) and most of the recharged water should be recovered from the aquifer using adequately placed water interceptors (NRC, 1994). Furthermore, pertinent information on water quality parameters and fate of various contaminants during aquifer passage is required for assessment and design of managed aquifer recharge (MAR) system used for recycled water treatment (Patterson et al., 2010). This information helps the planners to evaluate feasibility of SAT technology during pre-design, design and operation and maintenance phases.

2.2 SAT PRE-DESIGN CONSIDERATIONS

2.2.1 Intended use of SAT reclaimed water

Due to its resilience, SAT is used to treat a wide spectrum of wastewater effluents based on the intended use of the filtrate (Sharma et al., 2012). SAT can be used in combination with a conventional treatment technology as a polishing stage for the effluent, or to replace any specific stage(s) of treatment process. SAT reclaimed water suites a large variety of applications including landscape irrigation, residential, recreational, groundwater recharge, aquaculture and industrial cooling water (Huertas et al., 2008). Indirect potable reuse after SAT is also common (Fox et al., 2001a). Moreover, SAT percolate can be used for aquifer recharge to protect coastal aquifers against saline water intrusion and aquifer storage of surplus water for subsequent use in times of water shortage (Bouwer, 2002; Dillon et al., 2009). Furthermore, water reuse is promoted as a means of limiting arbitrary wastewater discharges to aquatic environments (Huertas et al., 2008) to protect in-stream and downstream users of that water against unacceptable pollution (Bouwer, 2002). Different reuse applications have different regulations and water quality and treatment process requirements (USEPA, 2006). Hence, it is critical at the planning stage to start with identifying what reuse application is needed. This will enable planners and designers to assess the feasibility of SAT with regard to the reuse purpose and pre- or post-treatment requirements to comply with the intended reuse water quality.

2.2.2 Public health

Protection of public health is the most critical objective in any water reuse program followed by preventing environmental degradation (USEPA, 2012). Groundwater recharge with reclaimed municipal wastewater effluent presents a prime health challenge that must be carefully evaluated prior to undertaking a recharge project (Asano and Cotruvo, 2004). Presence of pathogenic organisms in wastewater effluents and potential transmission of infectious diseases by these organisms is the centre of this concern (Metcalf et al., 2007; Vigneswaran and Sundaravadivel, 2004). In order to alleviate negative health impacts in any water reuse project, issues like proximity of human habitation to reuse site, human contact with the water, human ingestion of aerosol and direct exposure of wastewater to workers skin need to be considered (Toze, 1997). High removal efficiencies of contaminants can be achieved during SAT under optimum operating conditions with respect to travel time/travel distance, hydraulic loading rate and redox conditions (Sharma et al., 2012). Removal of organic compounds, nitrogen, phosphorus, suspended solids, bacteria and viruses in soil infiltration systems (i.e. SAT) is achieved through sorption, chemical reaction, biotranformation, die-off and predation (Kanarek and Michail, 1996). However, even advanced technologies for wastewater treatment (nutrients removal, reverse osmosis, activated carbon) suffer from a lack of scientific information on health effects when treated wastewater is reused to augment potable supplies (Westerhoff and Pinney, 2000). In general, to address the public health and environmental concerns related to wastewater reuse schemes, it is of a paramount importance to know the constituents present in a wastewater source and the level of treatment required to reduce these constituents to acceptable levels (USEPA, 2012).

2.2.3 Economic aspects

Technically, wastewater can be treated to any intended quality level. However, the price of the treatment influences the desired water quality and a compromise must then be reached between the quality and the cost at which such water quality could be achieved (Salgot, 2008). Wastewater reuse can help to maximize the use of limited water resources and contribute to economical development (Janosova et al., 2006) through reduction of budgets allocated for energy, chemicals procurement and reduction of highly treated water usage for non-potable purposes. Since water reclamation and reuse are not for free, it is prudent to identify the cost bearer (Salgot, 2008) during the planning phase and the potential treatment level that could be achieved at such a cost. This assessment helps pioneer proponents and experts to ensure project sustainability during the operation phase. In general however, the cost of SAT is relatively lower than that of conventional above-ground-treatment system and its operation is simple and no chemical or expensive treatment units and equipment are required (Sharma et al., 2012). Furthermore, reclaimed water is always perceived as a low-cost new water source during planning phase of water reclamation and reuse project. This assumption is true if water reclamation facility is situated near large agricultural or industrial beneficiaries and when reclaimed water does not require an additional treatment beyond the one from which the reclaimed water is delivered (Asano, 2002). Wastewater reuse in agriculture is considered as an integral element in sustainable management of limited freshwater resources. It provides potential economic and environmental benefits (Janosova et al., 2006) including consumption of less synthetic fertilizers and protection of water receptors from direct discharge of poor quality wastewater.

According to Asano (2002), construction cost breakdown for wastewater treatment plant in California, United States of America (USA) that treats 3,785 m^3/d up to secondary effluent level with total capital cost (USD $0.5/m^3$) is distributed as: primary treatment 24%, secondary treatment 40%, sludge treatment 22%, and control, laboratory and maintenance buildings 14%. Nema *et al.* (2001) postulated that cost savings up to 30% could be achieved by applying primary treated effluent (bypassing the secondary treatment) to the land. Furthermore, (Bouwer, 1991) reported the cost of SAT system to be less than 40% of equivalent above- ground-treatment. In the Shafdan, Israel, SAT field experience, a typical capital cost of 0.23-0.25 € is realized for every cubic meter of treated water with an operation and maintenance costs of 0.10-0.15 €/m^3 (Aharoni et al., 2011; Sharma et al., 2012). This capital cost includes the infiltration field, excavation equipment, sand replacement, pipelines, electro-mechanical parts, valves and pumps but, excludes costs of storage and distribution.

Even though costs associated with development of new SAT scheme might be relatively lower compared to conventional above ground treatment system, a market assessment for reuse potential is another necessary requirement during the planning stage (Al Kubati, 2013). This is because success of a water reuse scheme such as SAT is also dependent on guaranteeing markets for reclaimed water and therefore it is essential to find potential customers at this early stage (Tchobanoglous et al., 2003). While development of new water reuse projects should be based on investment cost recovery, current pervasive use of "free" raw wastewater in agriculture may adversely affect the willingness of farmers to pay for reclaimed water in the future.

2.2.4   Regulations and guidelines

Water reclamation and reuse guidelines and regulations are important to safeguard public health and limit adverse environmental impacts (Metcalf and Eddy, 2007). While regulations are legally adopted enforceable and mandatory, guidelines are advisory, voluntary and non-enforceable (Metcalf et al., 2007). Due to site specific nature of water reclamation projects, water reuse regulations and guidelines that specify the use of reclaimed water in various applications (i.e. irrigation, non-potable, industrial, recharge etc.) differ widely around the world. Furthermore, health risk perception pertaining to wastewater reuse varies significantly among various cultures across the globe. Pervasive reuse of raw wastewater in agriculture in some communities in developing countries epitomizes lack of investment costs to provide the appropriate technology, regulations and public awareness about health risks associated with the use of this type of low quality water. On the contrary, technology availability and presence of regulatory agencies to enforce appropriate regulations have enabled developed countries to meet their stringent reuse standards (Khouri et al., 1994). However, some countries and regions have imposed stringent set of standards similar to those in Australia and the U.S., while others have based their standards on the WHO guidelines for wastewater reuse which are less stringent as compared to Australia and the USA (Crook et al., 2005). An example of regional regulations is the state of California which has a long history of reuse. This state developed the first water reuse regulations in the USA since 1918 which has served as the basis for reuse standards development in other states in the U.S. and other countries (Crook et al., 2005).

2.2.5 Technical aspects

The performance of SAT system is affected by pre-treatment level of the wastewater effluent, site characteristics and operating conditions (Fox et al., 2001a). Pre-treatment of wastewater effluent directly affects organic and inorganic content of SAT feed water which in turn affects its operation and maintenance. On the other hand, site characteristics influence the suitability of potential site for SAT. Site characteristics are a function of local geology and hydrogeology (Fox et al., 2001a). The presence of an unconfined aquifer, an uncontaminated vadose zone with no restricting layer (such as clay lenses) coupled with soils that are coarse enough to give high infiltration and at the same time fine enough to provide good treatment are the site requirements for SAT infiltration basins (Bouwer, 1991). Furthermore, SAT sites are operated at alternating wet/dry cycles to restore infiltration rates and disrupt insects' life cycles (Fox et al., 2001a). To ensure continuous operation of SAT spreading basins especially during maintenance periods, the spreading basins can be subdivided into an organized system of smaller basins (USEPA, 2012). This will enable alternate filling and drying of the basin without flow disruption.

2.2.6 Socio-political aspects

Water shortage has triggered implementation of indirect and direct wastewater reuse projects in many developed and developing countries such as Singapore, Israel, Namibia, the USA, Australia and many European countries. Windhoek, Namibia commenced the direct wastewater effluent reuse in 1968. On the contrary, wastewater reuse practices in states of California, Texas and Virginia in USA are dominated by indirect reuse pratices (Po et al., 2003).

Aesthetics and public acceptance are of paramount importance in water reuse, especially where the public is directly affected by water reuse (Levine and Asano, 2004). Recharge systems (i.e. SAT) make the water reuse possible where religious taboos prevail against direct uses of "unclean" water (Warner, 2000). The use of natural systems for water reclamation enhances public confidence in water recycling projects that involve putting the water back into streams and aquifers before recovery for reuse (Dillon et al., 2006). One of the outstanding merits of SAT is that it breaks the pipe-to-pipe connection of directly reusing treated wastewater from treatment plant (Bdour et al., 2009). SAT makes potable-water reuse of the recycled water aesthetically more acceptable to the public (Bouwer et al., 2002) since its treated water comes from wells and not WWTP (Bouwer, 2002).

Public acceptability plays a pivotal role in implementation of wastewater reuse projects. According to Po et al. (2003), excrement, urine, saliva, dirt and mud are considered as general objects that can evoke public disgust. In addition to that, people may still perceive recycled water as disgusting regardless to the extent of advanced water treatment the recycled water underwent. Public perception on wastewater reuse is influenced by degree of human health protection, environment, treatment, distribution, conservation of socio-cultural makeup of the people involved (Bruvold, 1988; Lawrence et al., 2003). Frewer et al. (1998) concluded that people may consider recycled water too risky for some reasons such as the source of this water is not natural, the water may be harmful to people, fear from unknown future consequences, fear of not being unable to abundantly reuse water in the future and lack of trust and reliability on recycled water quality. Acceptability of wastewater reuse project is influenced by its proximity to people. The closer the reuse project to the people, the more it is opposed (Po et al., 2003). Nancarrow et al. (2002) conducted a study on public acceptability for recycled water. The respondents stated their consent with the water regardless to its source provided that it is safe and treated to comply with appropriate health standards. Nevertheless, consumers tend to pay less tariff for recycled water as they assume it to be of less quality (Po et al., 2003).

Consumers' involvement during the first stages of any reuse project is prerequisite for its long term public acceptability and success. According to Po et al. (2003), the purpose of community involvement should not be used as a ground to persuade or sell the use of recycled water to the community, but the authorities and process should be rather honest and transparent. Furthermore, precise and up-to-date information on the project should be frequently availed to the public, media, and educators. The public should be informed of the location of wastewater reuse fields and advised to prevent their children from entering these fields. Warning signs should be provided along reuse fields' borders (Lawrence et al., 2003).

Public acceptance is critical and therefore education and community programme must be arranged to assure the public of the purity and safety of reclaimed water (Bouwer, 1991). Factors that led to success of some existing water reuse projects in Australia, Singapore and the USA (Po et al., 2003) include:

- Long term commitment to inform and educate the local community about efficient water use and reuse.
- Careful planning with great emphasis on public involvement.
- Commitment to listen and address public or stakeholder concerns.
- Conducting studies by experts to convince both public and other stakeholders that the project does not pose any negative impacts to both the public and the environment.
- Gauging public acceptance by conducting comprehensive research project in order to better understand the public willingness to use reclaimed water.
- Public outreach through brochures and related fact sheets, video presentations about the project, feature stories in newspapers, and other media outlets.
- Gaining supports from different groups such as the environmental groups.
- Extensive educational programme using school presentations, tours to treatment plants, project exhibitions at local community events, and providing pamphlets to consumers with their water bills outlining the project.

The benefits that can be derived by involving public at the planning stage of a water reuse scheme include the following (Metcalf et al., 2007):

- Satisfying community water demands.
- Gaining public support and involvement in the project.
- Developing a broad market for reclaimed water.
- Improving project implementation.
- Establishing a two-way communication channel which eventually informs both the public and the planners about issues that may have been missed out or misunderstood by the other group.

Nevertheless, no formal standard framework has been developed to be used for achieving successful public involvement in a water reuse project.

2.2.7 Institutional aspects

Success of a water reuse project does not only depend on effectiveness and suitability of the technology adopted, but also on existence of an institutional framework under which distribution and safe use of the treated water can be efficiently achieved (Lawrence et al., 2003). Due to the multi-sectoral nature of water reuse projects, interests of various authorities, agencies, sectors and organizations involved should be considered and reconciled to achieve a successful project operation (Khouri et al., 1994). This entails that the governmental organizations involved in treated wastewater reuse should be clearly defined and their responsibilities should be clearly delineated (Lawrence et al., 2003) to avoid overlapping responsibilities and conflict of interest. A clearly spelled out institutional framework showing roles and responsibilities of each actor should be availed (Al Kubati, 2013). However, it is quite delicate to properly identify the stakeholders and institutions involved in reuse projects (Khouri et al., 1994).

As a first step towards a successful reuse scheme implementation, a government should be able to evaluate its institutional capability and flexibility in adopting the new framework in terms of stakeholders commitments, number and skills of staff as well as the financial investment required for restructuring and implementing the scheme. Provisions should be made to adequately staff and resource organizations charged with the responsibility for assessing, implementing, operating and monitoring effluent use schemes and enforcing compliance with regulations (Pescod, 1992). Institutional capacity and enforcement capabilities must be increased in most developing countries if wastewater reuse projects are to succeed (Khouri et al., 1994). Currently, lack of regulatory agencies or their incapability to enforce the adopted standards is a limiting factor in developing countries (Khouri et al., 1994).

2.3   SITE IDENTIFICATION AND INVESTIGATION

Site selection is a key element that determines the success of a SAT scheme as failure of SAT systems is most often related to improper or insufficient site evaluation (Crites et al., 2006; Reed et al., 1985). Factors like depth to groundwater table, groundwater flow patterns, redox conditions, soil characteristics, soil depth, proximity to conveyance channel and/or wastewater reclamation facilities are carefully checked and evaluated when selecting a suitable site for SAT system (Crites et al., 2006; Fox et al., 2001a; Harun, 2007). Furthermore, it is of economic importance to find a suitable site within a conveyable distance from the wastewater source (USEPA, 2006). A comprehensive list of the factors cited by Dillon et al. (2006) during the site selection process of the Alice spring SAT scheme in Australia included:

- Land availability
- Unconfined aquifer
- Depth of groundwater
- Presence of sufficiently permeable vadose zone
- Proximity to source water
- Proximity to locations of potential demand
- Preference to operate the site on the government owned land
- Avoidance of densely populated areas and avoidance of flood-prone land.

The first step involves gathering available data of potential sites to compare and evaluate through desk study, while the second step involves site data verification through further field investigation. To identify the potential land treatment sites, it is necessary to obtain data on land use, soil types, topography, geology, groundwater, surface water hydrology and applicable water rights issues (Crites et al., 2000; Crites et al., 2006; USEPA, 2006). Therefore, the important site specific factors that must be carefully analyzed are grouped under physical, hydrological, land use and economical factors. The following step is to conduct a thorough site investigation and laboratory testing.

2.3.1 Physical factors

The physical factors considered in this study are land availability, site grade (topography) and site susceptibility to flooding.

2.3.1.1 Land availability

Typical SAT systems by spreading basin method require a significant land surface area to allow infiltration of the wastewater into the aquifer and its subsequent treatment. Compared with other land treatment systems in which other secondary benefits can be derived from the land (i.e. crop or forest production, habitat enhancement, etc.), SAT requires long term commitment of the land area for treatment, with minimal secondary benefits (Metcalf et al., 2007). Basin size may range from less than 0.4 to more than 8 ha, and it is necessary to include at least 2 separate basins for even the smallest of systems (Crites et al., 2000; Metcalf et al., 2007). Normally apart from the basin area other additional land space is required to cater for buffer area (to screen SAT field from the public), on-site pre-treatment space, access roads, basin side slopes, berms and future expansion. An estimated land area required for treatment of primary and secondary effluent with a daily flow of 1 mgd (3785 m³/day) at the planning stage is shown in Table. According to Aharoni and Cikurel (2006), the area required for planning the whole SAT system including infrastructure is approximately double the area of the infiltration basins. Table 2.1 presents land requirements for primary and secondary effluent.

Table 2.1 SAT land area requirements based on effluent type

Type of wastewater effluent	Land requirement* (ha/m³.d)
Primary	0.0032
Secondary	0.0016

*Areas include an additional 20 % to cater for unused space. Source: (Crites et al., 2006; USEPA, 2006)

2.3.1.2 Topography

Topography is the variations in the elevation and steepness of the land surface that form the various components of the landscape. Movement of water within the ground is dictated by elevation difference of the site. Sites with high slopes or grades are of great concern, because this will require extensive cut-and-fill or related earthmoving operations which are not desirable for basin construction. According to Crites et al. (2006), sites that have extremely non-uniform topography do not absolutely preclude development of a SAT system, but rather result in the following:

- A significant increase in cost and complexity of site investigation.
- An increase in the cost of site preparation as a result of extensive cut-and-fill.
- Require heavy earth work machinery that can alter the necessary soil characteristics through compaction.
- Unstable soil conditions due to saturation of steep slopes.

The site selection criteria with respect to site grade are listed in Table 2.2.

Table 2.2 Suitability of site grades for SAT application

Parameter	Source		
	(USEPA, 1981)	(Crites et al., 2000)	(Crites et al., 2006)
Site grade (%)	15-20 (NS)	>20 (NS)	10-15 (NS)
	10-15 (LS)	12-20 (LS)	5-10 (S)
	5-10 (S)	0-12 (HS)	0-5 (HS)
	0-5 (HS)	–	–

HS: high suitability; S: suitable; LS: low suitability; NS: not suitable.
Based on these findings, slopes suitable for SAT basin construction should be less than 15 %, with sites having a slope of 0 to 5 % being the most suitable.

2.3.1.3 Susceptibility to flooding

SAT basins susceptible to flooding should be protected such that the loading cycles are not disrupted. Moreover, wastewater effluent applied onto SAT basins can be washed off by flooding. Therefore, flood hazard of each site should be evaluated in terms of the possible severity and frequency of flooding as well as the areal extent of flooding (Crites et al., 2000). Site susceptibility to flooding can affect its desirability, and therefore sites unsusceptible to flooding are recommended for basin construction of SAT systems.

2.3.2 Hydrogeological factors

SAT systems require unconfined aquifers, vadose zones free of restricting layers and soils that are coarse enough to allow for sufficient infiltration rates but fine enough to provide adequate filtration (Bouwer, 1987; USEPA, 2012). Hydrogeological parameters that determine the performance and applicability of SAT include: depth of vadose zone, soil type, type of aquifer and subsurface soil profile.

2.3.2.1 Depth of vadose zone

Surface spreading using infiltration basins is limited to unconfined aquifers with a vadose zone. This zone is the space between ground surface and the groundwater table including the capillary fringe (Metcalf et al., 2007), where most contaminants are removed. While groundwater table should be deep enough to facilitate purification of the infiltrating water and avoid mounding, it should be shallow enough to facilitate good recovery of the injected wastewater effluent (Akber et al., 2003). A minimum vadose zone thickness of 1-2 m is necessary to achieve sufficient removal of contaminants. Nevertheless, taking into account a minimum excavation depth of 1.5 m from ground level and a minimum mound rise of 1.5 m above the groundwater level, then the minimum thickness of vadose zone required when evaluating suitable sites should be 5 m.

2.3.2.2 Soil type

Soil can be classified based on the relative amounts of clay, silt and sand that it constitutes. This classification can be achieved using the soil-textural triangle shown in Figure 2.1. The degree of wastewater effluent treatment within the vadose zone is site specific and largely depends on the soil properties (composition and structure). Not all soils are appropriate for pollutants removal (Ho et al., 1992). Soil used for SAT systems should possess the suitable physical and chemical properties to achieve sufficient contaminants removal. Soils with substantial clay fractions have to be avoided since they could be impermeable which in turn increase land requirements for SAT percolation ponds. On the other hand, coarse sands yield high infiltration capacity, but remove less pollutants (Ho et al., 1992). The best surface soils for SAT systems are in the fine sand, loamy sand and sandy loam range (Crites et al., 2000; Pescod, 1992).

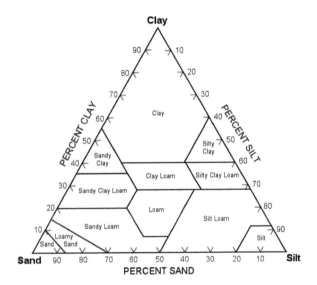

Figure 2.1 Triangular chart showing the percentages of sand, silt and clay in the basic soil-texture classes
Source: (Bouwer, 2002)

2.3.2.3 Permeability

Soil permeability is the ability of water to flow through the media. High permeability is amongst the various critical subsurface conditions necessary for SAT systems. This does not only allow for high turnover of wastewater effluent, but most importantly it reduces the size of infiltration area required for a given effluent flow rate. If a sufficiently permeable material is overlain by a low permeability top soil (i.e. about 1 m thick), then this top soil can be excavated such that the infiltration basin bottom is within the permeable material (Bouwer, 1999). Site selection criteria with respect to permeability of the most restrictive layers within the vadose zone as mentioned in several literatures is shown in Table 2.3.

Table 2.3 Permeability of most restrictive soil layer and its relative suitability for SAT

Permeability of most restrictive soil layer		Suitability
$(cm/hr)^{1,2}$	$(cm/hr)^{3}$	
0.15-0.5	–	Not suitable
0.5-1.5	–	Low
1.5-5.1	>2	Suitable
>5.1	–	High

Source: [1]Crites et al., 2000; [2]Crites et al., 2006; [3]USEPA, 1981

2.3.2.4 Type of aquifer

The aquifer in which SAT by surface spreading is intended has to be unconfined such that the applied effluent can infiltrate naturally through the vadose zone and mix with groundwater. The subsurface soil profile has to be free of restricting layers which impede vertical flow of water. Presence of clay lenses for example can render a site inappropriate for recharge applications (Houston et al., 1999).

2.3.3 Land use and location of SAT site

2.3.3.1 Land use

Land is developed and utilized for a broad range of purposes. Four categories of land use (Table 2.4) are considered in this study with respect to suitability for SAT system. The type of land use for a proposed SAT site influences project's feasibility and its suitability. Therefore, it is important to assess the land use and availability of reuse market at the vicinity of new SAT schemes to ensure their cost effectiveness throughout their life cycle.

Table 2.4 Effect of land use on suitability of a site for SAT

Land use category	Description
Agriculture or open space areas	Highly suitable for SAT application since cost of land will be relatively low. Locating a SAT system close to agricultural lands makes reclaimed water distribution economically feasible since reclaimed water is used for irrigation. Moreover, diverting the wastewater effluent for irrigation reuse is an advantage when only a portion of the effluent is intended for SAT or when the effluent volume is higher than what the SAT basins can accommodate.
Low density residential area	Less suitable for SAT compared to agricultural areas. Cost of land in locations allocated for residential purposes is relatively higher than that allocated for agricultural purposes. Land may be available, but depending on growth rate and expansion. This could deter future expansion areas for SAT might be difficult.
Residential urban	orHigh density areas are common to most cities, and land cost is very expensive. In these cases when SAT is desired, suitable sites may only be available at greater distances outside the city area. Therefore the possibility of finding a large area of land for SAT within an economically conveyable distance of the wastewater becomes difficult.
Industrial area	May be expensive, except for SAT site developed for industrial reuse of reclaimed water. Proximity of SAT site to location of demand will be regarded as economical. However, the possibility of lands being contaminated within this region is high.

2.3.3.2 Location of SAT site

The factors considered are the distance and elevation difference between SAT site and wastewater effluent source (i.e. WWTP). Distance from the WWTP to the SAT site and elevation difference between these two locations play an important role in the cost effectiveness of a SAT scheme. Pumping treated water over long distances and large elevations has a significant impact on operation and maintenance costs. Most costs associated with SAT water reuse system are ascribed to pumping the water from the recovery wells (Bdour et al., 2009). The cost of infrastructure (i.e. pipelines and possibly booster pumps) is directly proportional to the distance and elevation between the WWTP and the SAT site. Furthermore, conveyance and distribution of reclaimed water constitute substantial cost fraction of water reuse projects (Asano, 2002). Distance and elevation are important consideration during the site selection stages. The closer a SAT site is to a WWTP, the lesser will be the cost incurred in laying out a conveyance system and its subsequent maintenance (Akber et al., 2003). In evaluating these two parameters for a set of selected sites, it is sometimes of great importance to also compare the cost of pumping to a nearby site with the cost of conveyance by gravity to a far away site (USEPA, 2006).

The site selection criteria with respect to distance and elevation difference between the wastewater effluent source and the SAT site as mentioned in several literatures is shown in Table 2.5.

Table 2.5 Distance and elevation difference of the SAT site relative to the wastewater effluent source

Distance from wastewater source (km)	Elevation of SAT site relative to wastewater source (m)	Site suitability
0-3	<0	High
3-8	0-15	Suitable
8-16	15-60	Moderate
>16	>60	Low

Source: (Crites et al., 2000; Crites et al., 2006; USEPA, 1984)

2.3.4 Site investigation

SAT sites can be quite heterogeneous in nature with varying geology and hydrogeology. Such variability can occur within few meters (Al Kubati, 2013). This is why most engineering projects that have major components based on soil strata require detailed understanding of the proposed site geology (Sara, 2003). Field investigation is the second phase of an artificial groundwater recharge scheme that precedes the design phase and follows the preliminary phase which includes data collection, assessment of regulatory, legal, political, economic feasibility and conceptual planning (NRC, 2008). The investigation phase of SAT project is necessary because it requires a detailed understanding of the site subsurface geology and hydrogeology to determine site suitability for design. Hence, once a site has been identified or the most suitable site has been selected, the next step would be to physically explore the site and conduct some subsurface investigation works (USEPA, 1984; USEPA, 2006). This will help to verify the existing data and also identify probable or possible site limitations (Crites et al., 2000; Crites et al., 2006).

Filed testing for SAT site is always recommended to be conducted on actual basin location and at actual depth to avoid extrapolation of data from nearby sites (USEPA, 1984). Soil physical and chemical properties, actual permeability, absence of polluted areas and soil stratification (to determine depth of vadose zone and aquifer), groundwater level, quality and flow pattern, aquifer depth and conductivity are among the parameters frequently investigated (Crites et al., 2006; USEPA, 2006). The investigation helps to evaluate hydrogeology, lithology, depth to groundwater, confining zones, aquifer materials and aquifer properties (NRC, 2008). Drilling and construction of shallow and deep observation wells in the location will show the distribution of soil profiles in terms of sand and clay layers with their depth, thickness and composition. These data are essential for a better understanding of the unsaturated section as well as the aquifer properties and its capability to serve as a seasonal storage layer during SAT system operation (Aharoni et al., 2011). In order to estimate the land area needed for a certain volumetric recharge rate or the recharge rate that can be achieved with a certain land area, the actual site infiltration rates have to be determined by conducting "wet" infiltration tests (Bouwer, 2002). It is however difficult to decide the number of tests adequate for a site and their locations (Crites et al., 2000). Where extreme soil variability exists on soil maps, a large-scale pilot cell may be constructed to define site hydraulic characteristics (USEPA, 1981).

2.3.4.1 Test pits and boreholes

Site investigation works include both field and laboratory tests. Test pits are excavated with backhoes to inspect subsurface soil profile, texture, structure and to detect presence of any restricting (fine textured or cemented materials) layers within the vadose zone. Hydraulic conductivity of the restricting layers can also be tested at certain depths with infiltrometer tests or other methods (Bouwer, 1999). With the use of trenches or pits, it is possible to inspect soil profiles up to a depth of about 7 m (Bouwer, 1999). Additionally, groundwater is monitored if encountered within this excavation depth by installing piezometers to monitor seasonal level changes. To explore soils deposit below the limits of pit excavation up to groundwater and permeable layers, boreholes are made using augering or rotary drilling techniques (Reed et al., 1985). Soil samples are also collected during the excavation process for lab testing and analyzed for particle size distribution, cation exchange capacity (CEC), electrical conductivity (EC) and pH (Crites et al., 2000; USEPA, 2006). The results obtained are used to calculate sodium absorption ratio (SAR), exchangeable sodium percentage (ESP) and salinity (USEPA, 2012). Table 2.6 and Table 2.7 present, SAT site investigation requirements.

Table 2.6 Trial pits and boreholes investigation requirements for new SAT schemes

Type of investigation	Extent	Source		
		(Reed et al., 1985; USEPA, 1981)	(USEPA, 1984)	(Crites et al., 2000; USEPA, 2006)
Trial pits	No. of pits	Depends on site/project size soil uniformity. 3-5 per site	Minimum of 2 or 3 for even the smallest site	Depends on the site size. 8 ha site requires 6 to 10 pits
	Depth	≥ 3m	Up to 3m	2.4 - 3m
	Data required	Depth of profile, soil texture, soil structure, restricting layers and vertical conductivity.	Depth of profile, soil texture, soil structure, restricting layers and vertical conductivity.	Depth of profile, texture, structure, soil layers restricting percolation and vertical conductivity
Boreholes	No. of boreholes	Depend on soil uniformity and site size. Minimum of 3 per site.	Minimum 1 in every soil type or 1 for every 1-2 ha for large scale system (area up to 20 ha) with uniform site condition. 4-6 for small scale system (area > 5 ha) with uniform site conditions.	1 per 2 ha
	Depth	—	Penetrate to below the water table if it is 10-15 m and a few should extend through the whole saturated zone to determine aquifer thickness.	Penetrate below the water table if it is within 9 to 15 m
	Data required	Depth to groundwater, depth to impermeable layer(s) and horizontal conductivity.	—	Depth to groundwater, depth to impermeable layers and horizontal conductivity. Soil sampling for lab test.

Table 2.7 Detailed investigation requirements for infiltration and groundwater wells to develop new SAT scheme

Type of investigation	Extent	Source		
		(Reed et al., 1985; USEPA, 1981)	(USEPA, 1984)	(Crites et al., 2000; USEPA, 2006)
Infiltration and permeability	Type of test	Match the expected method of application if possible.	Preferably flooding basin test with minimum area of 7m². Cylinder infiltrometer or air entry permeameters (AEP).	Pilot-scale basin tests, at least 9.3 m² in area. Cylinder infiltrometer or AEP are other test methods. A combination of basin test and AEP, is recommended for most projects.
	No. of tests	Depends on size of site and uniformity of soil. Minimum of 2 per site.	Minimum of 1 for one major soil type, or 1 per 10ha for large area.	Depends on the system size and the uniformity of the soils and topography. 1 per site for small systems with uniform soils, or 1 for every 2–4 ha for large systems. For extremely variable site conditions, 1 full sized (0.4 to 1.3 ha) test basin will be required.
Groundwater wells	No. of wells	—	Minimum 3 (in the middle, up gradient and down gradient of the basin area near the project boundary).	Minimum 3 (in the middle, up gradient and down gradient of the basin area near the project boundary)
	Depth	—	Well bottom is usually between 3 - 10 m of the water table.	
	Data required	—	Depth to groundwater, thickness and permeability of the aquifer and groundwater quality.	Depth to groundwater, thickness and permeability of the aquifer and groundwater quality.

2.4 SAT DESIGN CONSIDERATIONS

2.4.1 Types of SAT systems

There are three main SAT systems commonly employed for wastewater effluent treatment including infiltration or spreading basins, vadose zone infiltration and direct injection or recharge wells (Bouwer, 1999; Metcalf et al., 2007). The key factors that determine which SAT system to adopt at the planning stage are: available information about soil, hydrogeology, land cost and wastewater pre-treatment requirements (Bouwer, 2002). While infiltration (recharge) basins are applicable where land is readily available and an unconfined aquifer with a vadose zone exists; vadose zone or direct injection wells may be used where these conditions are not favourable (Metcalf et al., 2007). This implies that the type of aquifer and land availability rank among the most critical factors that govern which SAT recharge method to adopt. Figure 2.2 presents different SAT systems used for water reclamation and aquifer recharge.

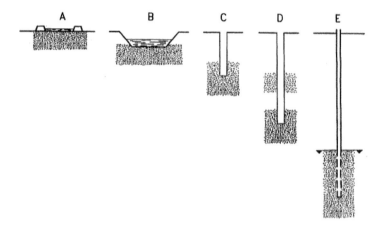

Figure 2.2 Types of SAT systems
Source: (Bouwer, 1999)

In infiltration basins, pre-treated wastewater is applied directly onto the surface of a wide land area, beneath which the treatment process occurs during vertical infiltration of the applied wastewater effluent. This SAT system requires high permeable soil within a reasonable depth of vadose zone that is not contaminated and does not contain layers of impermeable formations (Bouwer, 1999; Crites et al., 2000; USEPA, 2006). The aquifer below the vadose zone should be unconfined and sufficiently transmissive such that infiltrated water moves away from the recharge area without forming high groundwater mounds beneath the basin. Furthermore, the aquifer should be free of undesirable contaminants which may end up in undesirable locations (Bouwer, 1999). Where such conditions prevail, the wastewater effluent can be applied directly on the bare ground with berms (Figure 2.2 A) by the edges of the basin keeping the applied water in place. When a suitable soil exists beneath an overburden and within (an economical excavatable) reasonable depth (Figure 2.2 B), then the top layer can be excavated and the effluent applied into the basin (Bouwer, 1999). System A is termed surface spreading basin, while system B is known as excavated basin (Bouwer, 1999). When site topography calls for cut and fill to provide suitable basin levels, the basin surfaces should be located in cut sections and the excavated material can be placed and compacted in the berms (Crites et al., 2006).

In vadose zone systems, the effluent is applied via trenches or wells constructed within the vadose zone as shown in Figure 2.2 C and D, respectively. This system is applicable where land availability and/or top soil suitability restricts the use of surface spreading. Trenches are long and horizontal, typically less than 1 m wide and up to about 5 m deep, backfilled with coarse sand or fine gravel. (Bouwer, 2002). Vadose zone wells (also called recharge shafts or dry wells) are vertically constructed, typical wells vary in width from about 0.5 m up to 2 m in diameter and 30 to 46 m deep (USEPA, 2012). The wells are also backfilled with coarse sand or fine gravel and they offer a significant cost saving advantage over direct injection wells. Pre-treatment of the source water is critical to this type of system because they cannot be backwashed and a severely clogged well may be permanently destroyed (USEPA, 2012). The hydraulic capacity of this type of system is determined by the hydraulic conductivity of the vadose zone soils (NRC, 2008).

As the name implies, vadoze zone wells are used for direct recharge of the aquifer using reclaimed water of relatively high quality. They are used in situations where neither spreading basins nor vadose zone infiltration systems are applicable. These include when land area is limited, suitable soil throughout the aquifer depth is absent and aquifers are deep and confined (Bouwer, 2002). These wells are more expensive to construct compared to the vadose zone infiltration system. Nevertheless, they have a longer life span due to possibility of well cleaning and redevelopment (Metcalf et al., 2007; USEPA, 2012). Moreover, the use of this system minimizes risks associated with water loss since direct and quick recharge of target aquifer is possible (USEPA, 2012). The hydraulic capacity of direct recharge wells depends on the characteristics of the receiving aquifer (NRC, 2008). Types A and B are the most commonly used in SAT projects.

2.4.2 Pre-treatment of wastewater effluent

This is the additional treatment that the wastewater effluent receives before spreading onto basins. The aim of pre-treatment is to deliver treated effluent with the quality required either by regulations or for optimum operational purposes of SAT to meet reuse purposes. Pre-treatment can be applied to the effluent before conveyance to a SAT site, especially for long conveyance pipelines. This helps to prevent sedimentation and biofilms formation in the pipelines. SAT is flexible in terms of required wastewater effluent quality and pre-treatment requirements, and can be used for treatment of primary, secondary or tertiary effluents (Sharma et al., 2012). Pre-treatment levels vary from primary sedimentation to membrane filtration plus advanced oxidation. Table 2.8 shows various wastewater treatment steps and contaminants removed during each treatment step.

Table 2.8 Wastewater treatment level

Treatment level	Description
Primary	Removal of a portion of the suspended solids and organic matter from the wastewater.
Advanced primary	Enhanced removal of suspended solids and organic matter from the wastewater. Typically accomplished by addition of chemical addition or filtration.
Secondary	Removal of biodegradable organic matter (in solution or suspension) and suspended solids. Disinfection is also typically included in the definition of conventional secondary treatment.
Secondary with nutrient removal	Removal of biodegradable organics, suspended solids, and nutrients (nitrogen, phosphorus, or both nitrogen and phosphorus).
Tertiary	Removal of residual suspended solids (after secondary treatment), usually by granular media filtration or microscreens. Disinfection is also typically a part of tertiary treatment. Nutrient removal is often included in this definition.

Source: (Tchobanoglous et al., 2003)

The pre-treatment level required as stated above is mostly dictated by wastewater effluent quality and reuse regulations. In addition to this, the choice can also be related to operational or implementation problems, such as:

- When land is limited or expensive.
- When source water has high potential of forming clogging layer and this may result in a need for persistent basin cleaning.
- Where there is high potential for biofilms development in pipelines that convey source water to SAT site.
- When source water is aggressive and may lead to leaching or dissolution of heavy metals into groundwater or clogging of soil matrix.

Once any of the above problems is foreseen, then an appropriate pre-treatment should be chosen.

2.4.3 Hydraulic loading rate

This factor is essential in order to estimate land area needed for a certain volumetric recharge rate, or the recharge rate that could be achieved for a given area (Bouwer, 2002). The first step would be to perform a "wet" infiltration test to determine the infiltration rate of the soil (Bouwer, 2002). The design hydraulic loading rate (HLR) for SAT systems depends on the design infiltration rate and the treatment requirements (USEPA, 2006). Therefore, the HLR chosen for design would be the lesser value between the HLR which is a percentage of the field test infiltration rate and the loading rate based on treatment requirements (USEPA, 2006). According to Pescod (1992), these rates typically vary from 15 m/year to 100 m/year depending on soil, climate, quality of wastewater effluent, and frequency of basin cleaning. However, Bouwer (1999) postulated that these rates seem to be higher (30 m/year to 500 m/year) depending on the same factors. A more detailed breakdown of typical hydraulic loading rates for systems in relatively warm dry climates with good-quality input water is presented in Table 2.9.

Table 2.9 Hydraulic loading rates for different soil types

Type of soil employed	Hydraulic loading rate (m/year)
Fine textured soils like sandy loams	30
Loamy sands	100
Medium clean sands	300
Coarse clean sands	500

Source: (Bouwer et al., 2008)

However, infiltration rate depends on water viscosity which is a temperature dependent parameter (Jaynes, 1990). Besides, temperature affects biological activity which contributes significantly to formation of clogging layer and consequently reduction of infiltration rates (Le Bihan and Lessard, 2000).

HLR is usually the limiting design factor (LDF) when calculating the area needed for surface spreading, but in some cases the nitrogen or BOD loading may control the area needed (Crites et al., 2006). Hence, the area required based on the HLR is first calculated and then compared with the area requirement for BOD and nitrogen loading. The equations used in calculating the area requirement based on both the hydraulic loading rate and the nitrogen or BOD loading rate are shown below (Crites et al., 2006):

$$A = \frac{0.0001 \times Q\left(365\frac{d}{yr}\right)}{L_w} \tag{2.1}$$

Where A is the field area (ha), 0.0001 is used for metric conversion (ha to m^2), Q is the flow (m^3/d) and L is the hydraulic loading rate (m/yr).

The area obtained in this equation is then compared with the field area required based on nitrogen or organic loading rates, which is calculated as follows:

$$A = \frac{(8.34)CQ}{L} \tag{2.2}$$

Where A is the field area (ha), 8.34 is the conversion factor, C is the concentration of nitrogen or BOD (mg/L), Q is the flow (m³/d) and L is the limiting loading rate (kg/ha.d).

Infiltration basins have an extensive land requirements compared to other SAT methods such as the vadose zone well or direct recharge well. A hectare of recharge basins might be equivalent to a single recharge well (NRC, 2008).

2.4.4   Wetting and drying

Recharge basins during SAT are operated under alternating dry and wet cycles. Wetting and drying cycle is critical in design of SAT scheme since it is used to estimate the HLR and the number of basins. Clogging of the top surface soil during the basin flooding is a common phenomenon (Quanrud et al., 1996). Formation of clogging layer reduces infiltration rates of SAT systems to unacceptable level and the infiltration system must be desiccated to restore infiltration rates (NRC, 1994). The main aim of cyclic operation of SAT scheme is to increase infiltration rate, maximize nitrification and nitrogen removal (USEPA, 2006). While longer wet cycles increase the depths at which ammonia is adsorbed to the soil media, longer drying cycles increase aeration of the soil beneath the recharge basin by allowing oxygen to penetrate to greater depths. Oxygen is utilized by microorganisms to nitrify the adsorbed ammonium-nitrogen. A short wet period of less than 7 days is sufficient to prevent ammonium ion from breaking through sub-surface soils, whilst drying period should be long enough (greater/equal to 4 days for coarser soils) to enable oxygen to aerate the soil at deeper depths for subsequent utilization by nitrifiers to oxidize ammonium ions. However, Operating conditions must be based on local site characteristics and weather patterns since they are influenced by environmental factors including temperature, precipitation and solar incidence (Fox et al., 2001b).

2.4.5   Spreading basin design and layout

Basin design includes basin shape, geometry and arrangement. Choosing the suitable basin design would help to optimize operation, reduce groundwater mounding, utilize space or beautify the landscape. Therefore, depending on the type of basin function, its shape can either be triangular, square, rectangular, round, oval, or in a free form (Bouwer et al., 2008). Topography also plays a role when determining the shapes, arrangement and/or distribution of basins within a selected site.

Depth of basin should take into consideration maximum wastewater depth above the surface and the free board above this depth. Maximum wetting depth should not exceed 0.3 m for optimum infiltration rates and faster rate of wastewater turnover. To account for emergencies and where initial infiltration is slower than expected, basin depth should be at least 0.3 m deeper than the maximum design wastewater wetting depth (Crites et al., 2000). Another important design requirement is the minimum distance between the basin bottom and the groundwater, which is recommended to be no less than 1.5 to 3 m within 2 to 3 days of wetting (USEPA, 2006).

The area of recharge basins can also vary within the same project, since topography plays a role in sizing. Depending on project size, recharge area can vary from 0.2 to 2.0 ha for small to medium size project and 2.0 to 8.0 ha is the optimum area for larger projects. Three basins seem to be the minimum required. However, the higher the number of basins in a given area, the more flexible operation and maintenance are. The number of basins can be calculated based on the wetting and drying cycle and number of basins per set (i.e. basins fed at the same time).

$$TN_B = \frac{(W + D)}{W} \times N_B \qquad (2.3)$$

Where TN_B is the total number of basins on site, W is wetting period (days), D is drying period (days) and N_B is number of basins per set.

2.4.6 Groundwater mound

Groundwater mound occurs during interaction between infiltrated wastewater effluent and shallow native groundwater which leads to the rise of the latter. This is a usual problem where groundwater levels are shallow or when perched water develops as a result of shallow restricting layers (Metcalf et al., 2007). Excessive mounding slows down system's infiltration and reduces the effectiveness of treatment (USEPA, 1981). Moreover, undesired rise in groundwater can affect third party interests, such as basements, cemeteries, gravel pits, pipelines, old trees and low areas (Bouwer et al., 2008). Therefore, the potential of groundwater mounding is amongst the critical design requirements that should be probed before the system is designed and built if the geologic and hydrologic information is available for analysis (Crites et al., 2000).

According to Bouwer (1999), rise in mound can be reduced by pumping more groundwater in recharge areas having longer and narrower recharge areas or by reducing recharge rates. Besides, design recommendations propose that the horizontal separation between basin floor and groundwater should be 1.5 to 3 m within 2 to 3 days following a wetting period to avoid mound effect (USEPA, 1981; USEPA, 2006). However, Pescod (1992) postulated that the distance between groundwater tables and bottom of infiltration basins during wetting should be at least 1 m. This variation in horizontal distance in the literature seemingly depends on site characteristics such as soil type, texture and the depth of the vadose zone.

2.4.7 Abstraction and monitoring wells

Abstraction system normally consists of recovery wells strategically designed and situated around the infiltration basin to satisfy both hydraulic and water quality requirements. Recovery wells for pumping water after SAT from the aquifer can be located such that they pump nearly 100% reclaimed water or pump a mixture of reclaimed water and natural groundwater (Bouwer, 2002). In some schemes, recovery wells are located primarily for practical reasons such as proximity to point of use or conveyance system (NRC, 2008). This arrangement has the advantage of reducing project cost by eliminating the installation of conveyance pipelines and its subsequent maintenance costs (Al Kubati, 2013).

Wells may be located midway between two recharge areas or on either side of a single recharge strip or may surround a central infiltration area (USEPA, 2006). Locating the recovery wells at as far distance as possible from the spreading basins increases the flow path length and hydraulic residence time of the applied wastewater. These separations in space and time contribute to the assimilation of the treated wastewater with the other aquifer contents (Asano and Cotruvo, 2004). In order to achieve an adequate SAT treatment for wastewater effluent, the distance and transit time between infiltration basins and wells or drains should be as great as 50 to 100 m and perhaps 6 months of retention time (Asano and Cotruvo, 2004). The state of California guidelines for groundwater recharge state that recovery wells for potable reuse using SAT should be placed at least 300 m away from the basins and should pump a blend of at least 80% natural groundwater and not more than 20% reclaimed water (Bouwer et al., 2008).

As part of SAT system, monitoring wells are planned and situated between infiltration basins and abstraction wells. These wells must be capable of obtaining independent samples from each aquifer that potentially conveys the recharged water (NRC, 2008). Monitoring wells must be sampled for total organic carbon (TOC), total nitrogen, *total coliforms* and other constituents specified by regulations that are identified through reclaimed water monitoring (NRC, 2008). TOC is the most common monitoring parameter for gross measurement of organic content in reclaimed water used for potable purposes. It is used as a measure of treatment process effectiveness (Crook et al., 2005). The use of a non-specific chemical indicator such as TOC along with measurements and criteria for specific chemicals (e.g., benzene or nitrosamines) can show that a large portion of the organic chemicals of most types has been removed by the treatment technology and ensure that specific measurable hazardous chemicals do not exceed limits (Asano and Cotruvo, 2004).

2.4.8 Travel time

Most contaminants removal occurs in the first few meters of the vadose zone, but dilution with groundwater and residence in the aquifer are responsible for further reduction and removals especially of phosphorous, viruses and the more persistent micro-pollutants. Residence time or travel time plays a major role in design as it helps to predict water quality after SAT. The relative placement of both the monitoring and recovery wells relative to the infiltration basins is based on travel time. Dampening pollution incidence is another function of travel time. Some of the factors that influence or determine the residence time of applied effluent in the aquifer include: effluent quality (i.e. primary, secondary or tertiary), pre-treatment, reuse purpose and regulation requirements. Several reuse guidelines have specified residence time that the recharge water must stay in the aquifer before abstraction for reuse. However, these values are either based on "rule of thumb" or based on a typical pathogen (specifically virus) inactivation rates and do not consider other critical factors such as site-specific conditions. An example is the state of California groundwater recharge regulations that specifies a minimum residence time of one year for injected water and six month for surface spreading (NRC, 2008).

2.4.9   Post-treatment of the reclaimed water

Post-treatment or polishing stage is in some cases required for SAT water after a certain residence time prior to its distribution. The type and extent of post-treatment depends on the intended application of the SAT filtrate (Sharma et al., 2012). The choice of post-treatment is determined by regulations or guidelines. Post-treatment may also be required in order to prevent damage to water distribution systems or for public health and aesthetic reasons (NRC, 2008).

However, changes in operating conditions at SAT site may necessitate post-treatment of the abstracted water. In the case where anoxic conditions develop in the vadose zone at a SAT site leading to the solubilisation of reduced manganese (Mn), iron (Fe) and arsenic (As) from the aquifer materials, appropriate post-treatment will be required after recovery of the recharged groundwater (NRC, 2012). Experience in the Dan region, Israel, SAT field showed that after 20 years of operation, problem of high concentration of manganese appeared in areas close to the basins which led to changes in the recovery wells operational regime and closure of some wells (Aharoni et al., 2011; Goren, 2008). In such cases, conventional water treatment technologies such as aeration followed by rapid sand filtration (that will serve as post-treatment) will be necessary to address these problems. Examples of post-treatment technologies adopted at some SAT schemes around the world, are shown in Table 2.10.

Table 2.10 Examples of post-treatments at some SAT sites around the world

Project	Type of the effluent used	Post-treatment after recovery
Dan Region, Israel[1]	Secondary effluent	Intermediate chlorination
Dan Region, SWITCH project, Israel[1]	Secondary effluent	NF
Atlantis, South Africa[2]	Secondary effluent	Ion exchange + Chlorination
Torreele/St-Andre, Belgium[3]	Advanced treated effluent	Aearation + RSF+UV

Source: [1]Aharoni et al. (2011); [2]Tredoux et al. (2012); [3]van Houtte et al. (2012)

2.5   OPERATION, MAINTENANCE AND MONITORING OF SAT SYSTEMS

2.5.1   Operation and maintenance

Infiltration basins are the most widely used means of recharge due to their resilience in terms of space usage and low maintenance (Asano and Cotruvo, 2004). Operation of SAT is dominated by application of wastewater to infiltration basin at rates higher than evapotranspiration rate during a period of time termed wetting time (Abushbak, 2004). To achieve more resilient operation approach during SAT, more than one field (infiltration basin) is used to periodically dry one field and keep the other field in service. Drying periods range from two to three days for secondary effluent and one to two weeks for primary effluent (Bancolé et al., 2003). However, operating conditions must be based on local site characteristics and weather patterns since they are influenced by environmental factors including temperature, precipitation and solar incidence (Fox et al., 2001b).

During SAT operation, wastewater is applied to infiltration basin and left to infiltrate soil strata underlying the basin (Fox et al., 2001b). No pounding (increase in depth) of the applied water takes place until saturation conditions of the interface between wastewater and soil increase some orders of magnitude. Application of wastewater at rates higher than those of infiltration results in increase in (surface) water depth in the basin (Figure 2.3). Water depth is maintained constant (during flooding) by using a motorized valve activated by water level float which discharges the overflow (Abushbak, 2004).

Groundwater table starts to increase under the infiltration basin during the flooding cycle. On the other hand, formation of clogging layer at the bottom of the basin due to physical (filtration), chemical (precipitation of minerals) and biological (growth of micro-organisms and production of polysaccharides) processes (Bouwer et al., 2002), results in low hydraulic conductivity of the sediments beneath the pond (Kildsgaard and Engesgaard, 2001). At this stage, application of wastewater is ceased and pounding depth decreases gradually with time until the water disappears from the surface leading to the end of drainage period which marks the beginning of drying period (Abushbak, 2004). Consequently, groundwater table under the basin decreases until the next flooding cycle commences. Abushbak (2004) asserted that both drainage and drying periods are collectively known as drying cycle.

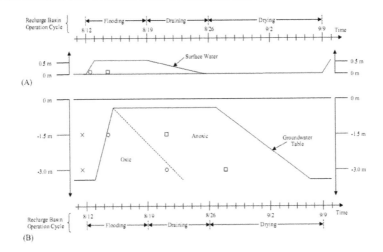

Figure 2.3 Typical infiltration basin operation (i.e. 1 week wetting and 3 weeks drying) cycles at Mesa water reclamation plant showing (A) water level in the basin and (B) groundwater level.
Source: Adapted from Montgomery-Brown *et al.* (2003)

Operation of SAT basin is influenced by the depth of wastewater effluent in infiltration basin. The average difference between water level and the bottom of infiltration basin is termed as water depth. Though high water head resulting from high water level could lead to high infiltration rates, it could eventually impede water percolation through clogging layer compression. Increasing water depth in the basin with low infiltration rates leads into longer exposure of unicellular algae (i.e. *Carteria klebsii*) to sunlight and aggravates their growth (Bouwer and Rice, 1989; NRC, 1994). Thus, shallow infiltration basins with water depths not greater than 0.5 m are generally favoured over deep basins (Bouwer et al., 2002) because the turnover rate of sewage applied to shallow basins is faster than for deep basins of the same infiltration rate, thus giving suspended algae less time to develop in shallow basins (Pescod, 1992).

Even though cyclic wetting and drying operation of an infiltration basin during SAT facilitates the recovery of infiltration rates, periodic removal of the clogging layer by scraping or racking is necessary for a successful long-term SAT operation.

2.5.2 Monitoring of SAT system

2.5.2.1 Wastewater effluent

Wastewater effluent is monitored at SAT site to assess WWTP performance, compare its quality with regulations requirements, determine if change in SAT operation is required and calculate organic carbon and nitrogen content of the effluent. A weekly or monthly monitoring measurement of BOD, DOC, TOC, TSS, NH_4-N, pathogens indicators, total phosphorus and heavy metals helps assess effluent quality (Idelovitch et al., 2003).

2.5.2.2 Depth of wastewater in spreading basin

In addition, to controlling wetting/drying cycles as mentioned above, maintaining the depth of wastewater effluent above the infiltration surface is also an operational requirement that reduces decline in infiltration rates. Wastewater effluent depth less than 0.3 m tends to keep the clogging layer loose and thereby it becomes more permeable compared to clogging layers that are subjected to higher water depth. Moreover, lower water depth in shallow basins have shorter drying periods, and thereby result in higher effluent turnover with insufficient periods for algae growth and insect life cycle. A staff or a graduated pole can serve as a measuring device through which the operator can monitor the depth. Where sensors or cameras are used instead, then this can be monitored from a control room. Increase in depth can be controlled by regulating influent flow or diverting flow to other basins. Also if basin colour starts turning green due to algae growth, flow should be diverted or stopped.

Based on this, it is recommended that the effluent level within the basin should be monitored on a daily bases. The level should not exceed 0.3 m. In other words flow can be controlled to ensure that this depth is not exceeded as follows:

$$Q \leq (IR_{avg} + 0.3) \times A_B \qquad (2.4)$$

Where Q is flow per day into the basin (m^3/day), IR $_{avg}$ is average infiltration rate of the basin (m/day), 0.3 is maximum effluent height in basin (m) and A$_B$ is basin area (m^2).

However, due to fall in infiltration rate towards the end of the wetting periods, especially when long periods are used, level will need to be monitored at least twice per day. This is based on the fact that infiltration rates as a result of clogging usually decrease by 50% (Metcalf et al., 2007).

2.5.2.3 Groundwater mound

Rise in groundwater level is monitored below or adjacent to infiltration basin. Groundwater level should be 1.5-3.0 m to avoid any interference with the basin bottom (USEPA, 2006). Regular monitoring is required after two or three days after starting the wetting cycle.

2.5.2.4 Groundwater quality

Groundwater quality data is necessary to evaluate the potential chemical reactions which may occur between the recharge water and the local groundwater when they mix in the aquifer (NRC, 2008). Many source waters have relatively higher redox potential (ORP) and dissolved oxygen concentrations compared with native water in deep portions of unconfined aquifers (USEPA, 2012). These source waters can react with the aquifer matrix (which is in equilibrium under reduced conditions) leading to a change in the stored and recovered water hydrogeochemistry (USEPA, 2012). Significant difference between chemistry of these waters is responsible for leaching of heavy metals. In an aquifer that contains reduced minerals such as arsenopyrite (reduced iron sulfide) or other reduced forms of arsenic minerals, an oxidising source water can oxidise these minerals and can cause the release of arsenic into the stored water (NRC, 2008). Reducing conditions can also arise in an aquifer when source water containing dissolved organic carbon recharges the aquifer which in turn leads to the release of iron and other metals and metalloids (including arsenic) into the groundwater (NRC, 2008). Methods that can be used to minimize leaching/transport of arsenic and other trace inorganic can include controlling the pH and matching the redox potential of the recharge water with the redox potential of the native groundwater (USEPA, 2012). Dilution on the other hand does improve the groundwater quality. An example of that is the Alice spring project where the native groundwater prior to SAT was brackish and unsuitable for irrigation of most crops. However, salinity improved and water became suitable for irrigation after implementation of SAT (Miotli ski et al., 2010).

In general, SAT system could operate for decades with marginal or no adverse effect on groundwater. After two decades of continuous SAT site operation and monitoring, no deleterious effect on groundwater was found in Los Angeles County, California, USA (Nellor et al., 1985).

2.5.2.5 Reclaimed water

Depending on intend of reuse, the frequency at which reclaimed water parameters are measured to decide on post-treatment requirement ranges from daily to monthly. Parameters like TDS, electrical conductivity (EC), pH, sodium absorption ratio (SAR), NO_3-N, NH_4-N and PO_4-P are crucial to be measured for agricultural reuse (Kandiah, 1990; Metcalf et al., 2007; Pescod, 1992). Nevertheless, industrial reuse requires monitoring of pH, BOD, TSS and *faecal coliforms* (Pescod, 1992).

2.6   OTHER CONSIDERATIONS

The use of reclaimed wastewater including groundwater recharge for a variety of applications has been implemented and it is safely undertaken provided that appropriate planning, treatment, water quality control, assessment and precautions are followed (Asano and Cotruvo, 2004). The successful implementation of a SAT project therefore does not end after a system is constructed and operated, but to a large extent also depends on system/activities put in place during the project life time, such as identifying benefits achieved, continuous monitoring and control programme and data gathering for research (Al Kubati, 2013).

2.6.1 Long-term impact of SAT systems

Even though renovated wastewater from the SAT process is of much better water quality than the influent wastewater, it could be of lower quality than the native groundwater (Asano and Cotruvo, 2004). In the cases where native groundwater is originally of a lower quality, SAT water can improve the groundwater quality. This has been the case in the Alice Spring, Australia SAT facility where the salinity of the groundwater improved after SAT and was regarded suitable for agriculture (Miotli ski et al., 2010). SAT system monitoring program is conducted through detection of any groundwater quality degradation that might be caused by infiltration of municipal wastewater effluent through SAT. Besides, remediation of any potential risk to ensure that the system is performing as designed. The design and management of SAT process should therefore aim to avoid wastewater encroachment into the native groundwater and to use only a portion of the aquifer (Asano and Cotruvo, 2004). One of the major concerns associated with the use of reclaimed water is the potential presence of low concentrations of a range of organic micropollutants. In addition, negative water quality changes can also occur when recharge water differs significantly in quality with the groundwater leading to leaching of iron or arsenic (Patterson et al., 2010). Even though some heavy metals and phosphorus are removed during SAT, it is not sustainable for long-term operation of SAT systems as the removal mechanism of these substances relies mainly on adsorption (Sharma et al., 2012).

Further research is needed in developed and developing countries to verify the reliability of water quality improvements within aquifer systems under a wider range of scenarios that would assist in identifying risks for use in localized risk assessments and enable indirect reuse of reclaimed water (Dillon, 2005). Monitoring the performance of the existing schemes to assess the degree to which they have achieved their stated objectives with reporting to successes and failures will be beneficial for future projects (Dillon, 2012).

Despite all field and research achievements in wastewater treatment and reuse, more effort is still needed to overcome the disgust and misconception caused by using misleading terminology. Terms like wastewater, direct water reuse, black water, grey water, yellow water and toilet-to-tap connection should be replaced with more relevant terms that do not negatively affect public perception and engender disgust.

2.6.2 Recreation and public environmental education

Ponds used for water reclamation are potential sites for public recreation activities and environmental education (Metcalf et al., 2007). Relatively deep recharge basins can be utilized for public recreation use. However, some of the problems that may arise include maintaining basin water level for boating and increased basin clogging due to rich aquatic environment (Bouwer et al., 2008). Besides, deep basins reduce hydraulic loading capacity because of the high water depth which compresses the clogging layer at the bottom of the basin (Bouwer et al., 2008). A solution to these problems if such deep basins (lakes or ponds) are desired is to have a mixture of engineered high infiltration basins and a couple of these low infiltration rate deep ponds for recreational purposes (Bouwer et al., 2008). Basins can be designed to support both wildlife and maintain a population of fish having a trough at the deep end of the basin can serve as a sanctuary place for fishes and other aquatic life forms during periods of drying (Metcalf et al., 2007). When ponds are designed for public attraction, they can have free shapes in order to blend with the landscaping and vegetation (Bouwer et al., 2008). The Santee Recreational Lakes project in San Diego, California is a good example where reclaimed water ponds were employed for public recreation use. The recreational reserve covers an area of 77 ha and hosts more than 550,000 visitors per year with activities such as fishing, boating and camping (Metcalf et al., 2007). The major attraction is the seven lakes system which covers an area of 24 ha. These lakes receive high quality water after SAT. This project generates sufficient fund that makes it completely self-reliant and has gained high level of public acceptance that has rendered it a popular recreational location. Educational potential of such systems can be increased by the use of visitor centres, descriptive signs and environmental education programs (Metcalf et al., 2007).

2.7 TECHNOLOGY TRANSFER TO DEVELOPING COUNTRIES

SAT is equally attractive for developed and developing countries as it removes multiple contaminants and minimizes the use of chemicals and energy (Sharma et al., 2012). Application of SAT technology in arid and semi-arid regions of the world where groundwater resources have been over exploited can augment water supply (Sharma et al., 2008). In many developing countries such as Ghana, Senegal, Pakistan, etc., wastewater is used without any treatment for crop, fodder, and green space irrigation with a high risk of easy transmission of waterborne diseases and groundwater pollution (Meric and Fatta, 2007). Development of SAT systems in these countries would ensure a safe reuse of reclaimed water for agricultural and other indirect potable reuse purposes. A successful transfer of technology can occur if the recipient is sufficiently capable of maintaining and fully utilizing the technology (Harun, 2007). Transfer of technology is more than just the moving of high-tech equipment from the developed to the developing world or within the developing world. It also includes total systems and their component parts including know-how, goods, services, equipment, organizational and managerial procedures. There is no single strategy for a successful knowledge transfer that is appropriate to all situations. The recipients will choose a technology which is appropriate to their actual needs, circumstances and capacities (UNEP, 2003). Since SAT is a low-tech reuse technology that does not require considerable operator expertise, the current knowledge accumulated at different SAT sites in the developed world could be easily transferred to the developing world.

In regions where SAT is already in practice, the experience gained and the numerous studies conducted can be very useful to facilitate replication of the systems on new sites. Nevertheless, more work is needed to develop a framework through which such replication could be easily executed. Water reclamation and reuse projects including

SAT are practiced in several states in USA with the state of California being the pioneer in this field. Furthermore, experience with SAT at one site in Israel and Australia is being replicated at other sites in these countries.

SAT has gained recognition as a cost effective, sustainable, simple and robust water reclamation and reuse technology. Its success has been documented in many parts of the developed world. However, many regions that face or likely to face water stress on their renewable source of freshwater have not exploited this technology to its maximum potential. Several reasons may be behind this slow identification and application of this technology in these regions. Current SAT experiences in the developed world are site specific while little awareness in developing countries might have deterred development and implementation of this technology. The difficulty of explaining and disseminating research-based information is further complicated by the public's current aversion to some forms of water recycling. Public outreach program addressing issues such as public participation, building public trust, information management and communication is a means of bridging the gap between institutions and end users. Websites, brochures, presentations, seminars, fieldtrips or excursions to existing SAT facilities and educating younger generations, are positive means of spreading awareness and information dissemination (Hartley, 2006).

2.8 REFERENCES

Abushbak, T. (2004). An Experimental and Modeling Investigation of Soil Aquifer Treatment System-Gaza City Case. PhD thesis, Royal Veterinary and Agricultural University, Copenhagen, Denmark.

Aharoni, A. and Cikurel, H. (2006). Mekorot's research activity in technological improvements for the production of unrestricted irrigation quality effluents. *Desalination,* **187**(1), 347-360.

Aharoni, A., Guttman, J. and Cikurel, H. (2011). Guidelines for Design, Operation and Maintenance of SAT (and hybrid SAT) Systems. [Online] http:www.switchurbanwater.eu/outputs/pdfs/W3-2_GEN_RPT_D3.2.1f-i.

Akber, A., Al-Awadi, E. and Rashid, T. (2003). Assessment of the use of soil aquifer treatment (SAT) technology in improving the quality of tertiary treated wastewater in Kuwait. *Emirates Journal for Engineering Research,* **8**(2), 25-31.

Al Kubati, K. M. A. (2013). Development of Framework for Site Selection, Design, Operation and Maintenance for Soil Aquifer Treatment (SAT) Systems. MSc Thesis MWI 2013-17, UNESCO-IHE, Delft, The Netherlands.

Asano, T. (2002). Water from(waste) water- the dependable water resource. *Water Science and Technology,* **45**(8), 24.

Asano, T. and Cotruvo, J. (2004). Groundwater recharge with reclaimed municipal wastewater: health and regulatory considerations. *Water Research,* **38**(8), 1941-1951.

Bancolé, A., Brissaud, F. and Gnagne, T. (2003). Oxidation processes and clogging in intermittent unsaturated infiltration. *Water Science and Technology,* **48**(11), 139-146.

Bdour, A., Hamdi, M. and Tarawneh, Z. (2009). Perspectives on sustainable wastewater treatment technologies and reuse options in the urban areas of the Mediterranean region. *Desalination, 237*(1-3), 162-174.

Bouwer, H. (2002). Artificial recharge of groundwater: Hydrogeology and engineering. *Hydrogeology Journal,* **10**(1), 121-142.

Bouwer, H. (1999). Artificial recharge of groundwater: Systems, design and management. *In Hydraulic Design Handbook,* ed. L. W. Mays, McGraw-Hill. New York, pp. 24.1-24.44.

Bouwer, H. (1987). Design and Management of Infiltration Basins for Artificial Recharge of Groundwater. *In 32nd Annual New Mexico Conference on Groundwater Management.* Albuquerque, NM.

Bouwer, H. (1991). Role of groundwater recharge in treatment and storage of wastewater for reuse. *Water Science and Technology,* **24**(9), 295-302.

Bouwer, H., Pyne, R., Brown, J., Germain, D., Morris, T., Brown, C., Dillon, P. and M., R. (2008). *Design, Operation, and Maintenance for Sustainable Underground Storage Facilities.* AWWA Research Foundation, IWA, USA.

Bouwer, H. and Rice, R. (1989). Effect of water depth in groundwater recharge basins on infiltration. *Journal of Irrigation and Drainage Engineering,* **115**, 556-567.

Bouwer, H., Rice, R., Lance, J. and Gilbert, R. (2002). Artificial recharge of groundwater: hydrogeology and engineering. *Hydrogeology Journal,* **10**(1), 121-142.

Bruvold, W. (1988). Public opinion on water reuse options. *Journal of Water Pollution Control Federation,* 45-49.

Crites, R., Reed, S. and Bastian, R. (2000). *Land Treatment Systems for Municipal and Industrial Wastes.* McGraw-Hill Professional.

Crites, R. W., Reed, S. C. and Middlebrooks, E. J. (2006). *Natural Wastewater Treatment Systems.* CRC Press, Boca Raton, Florida, USA, pp 413-426.

Crook, J., Mosher, P. and Casteline, J. (2005). *Status and Role of Water Reuse.* Global Water Research Coalition, London, UK.

Dillon, P. (2012). Chapter 17- General Design Considerations. *In Water Reclamation Technologies for Safe Managed Aquifer Recharge,* IWA Publishing. UK, pp. 299 - 310.

Dillon, P. (2005). Future management of aquifer recharge. *Hydrogeology Journal,* **13**(1), 313-316.

Dillon, P., Pavelic, P., Page, D., Beringen, H. and Ward, J. (2009). *Managed Aquifer Recharge: An Introduction.* National Water Commission, Australia.

Dillon, P., Pavelic, P., Toze, S., Rinck-Pfeiffer, S., Martin, R., Knapton, A. and Pidsley, D. (2006). Role of aquifer storage in water reuse. *Desalination,* **188**(1-3), 123-134.

Fox, P., Houston, S. and Westerhoff, P. (2001a). *Soil Aquifer Treatment for Sustainable Water Reuse.* American Water Works Association, Denver, Clorado, USA.

Fox, P., Houston, S., Westerhoff, P., Drewes, J., Nellor, M., Yanko, B., Baird, R., Rincon, M., Arnold, R. and Lansey, K. (2001b). An Investigation of Soil Aquifer Treatment for Sustainable Water Reuse. *Research Project Summary of the National Center for Sustainable Water Supply (NCSWS),* Tempe, Arizona, USA.

Frewer, L., Howard, C. and Shepherd, R. (1998). Understanding public attitudes to technology. *Journal of Risk Research,* **1**(3), 221-235.

Goren, O. (2008). Geochemical Evolution and Manganese Mobilization in Organic Enriched Water Recharging Calcareous-sandstone Aquifer; Clues from the Shafdan Sewage Treatment Plant, Hebrew University. Israel.

Hartley, T. W. (2006). Public perception and participation in water reuse. *Desalination,* **187**(1-3), 115-126.

Harun, C. M. (2007). Analysis of Removal of Multiple Contaminants During Soil Aquifer Treatment. MSc Thesis. MWI 2007-18, UNESCO-IHE. Delft, The Netherlands.

Ho, G., Gibbs, R., Mathew, K. and Parker, W. (1992). Groundwater recharge of sewage effluent through amended sand. *Water Research,* **26**(3), 285-293.

Houston, S. L., Duryea, P. D. and Hong, R. (1999). Infiltration considerations for ground-water recharge with waste effluent. *Journal of Irrigation and Drainage Engineering,* **125**(5), 264-272.

Huertas, E., Salgot, M., Hollender, J., Weber, S., Dott, W., Khan, S., Schäfer, A., Messalem, R., Bis, B., Aharoni, A. and Chikurel, H. (2008). Key objectives for water reuse concepts. *Desalination,* **218**(1-3), 120-131.

Idelovitch, E., Icekson-Tal, N., Avraham, O. and Michail, M. (2003). The long-term performance of Soil Aquifer Treatment(SAT) for effluent reuse. *Water Science and Technology: Water Supply,* **3**(4), 239-246.

Janosova, B., Miklankova, J., Hlavinek, P. and Wintgens, T. (2006). Drivers for wastewater reuse: regional analysis in the Czech Republic. *Desalination,* **187**(1-3), 103-114.

Jaynes, D. (1990). Temperature variations effect on field-measured infiltration. *Soil Science Society of America Journal,* **54**(2), 305.

Kanarek, A. and Michail, M. (1996). Groundwater recharge with municipal effluent: Dan region reclamation project, Israel. *Water Science and Technology,* **34**(11), 227-233.

Kandiah, A. (1990). Water Quality Management for Sustainable Agricultural Development. *Natural Resources Forum.* Wiley Online Library. 22-32.

Khouri, N., Kalbermatten, J. M. and Bartone, C. (1994). *The Reuse of Wastewater in Agriculture: A guide for planners.* UNDP-World Bank Water and Sanitation Program, World Bank. Washington DC, USA.

Kildsgaard, J. and Engesgaard, P. (2001). Numerical analysis of biological clogging in two-dimensional sand box experiments. *Journal of Contaminant Hydrology,* **50**(3-4), 261-285.

Lawrence, P., Adham, S. and Barrott, L. (2003). Ensuring water re-use projects succeed—institutional and technical issues for treated wastewater re-use. *Desalination,* **152**(1), 291-298.

Le Bihan, Y. and Lessard, P. (2000). Monitoring biofilter clogging: biochemical characteristics of the biomass. *Water Research,* **34**(17), 4284-4294.

Levine, A. D. and Asano, T. (2004). Peer reviewed: recovering sustainable water from wastewater. *Environmental Science and Technology,* **38**(11), 201-208.

Martin, J. and Koerner, R. (1984). The influence of vadose zone conditions in groundwater pollution: Part II: Fluid movement. *Journal of Hazardous Materials*, **9**(2), 181-207.

Meric, S. and Fatta, D. (2007). Wastewater reuse, risk assessment, decision-making - a three-ended narrative subject. Wastewater Reuse–Risk Assessment, *Decision-Making and Environmental Security*, 193-204.

Metcalf, E., Asano, T., Burton, F., Leverenz, H., Tsuchihashi, R. and Tchobanoglous, G. (2007). Water Reuse: Issues, Technologies, and Applications, Mc-Graw Hill. NewYork, USA.

Miotli ski, K., Barry, K., Dillon, P. and Breton, M. (2010). Alice Springs SAT Project Hydrological and Water Quality Monitoring Report 2008-2009. CSIRO. 1835-095X.

Montgomery-Brown, J., Drewes, J., Fox, P. and Reinhard, M. (2003). Behavior of alkylphenol polyethoxylate metabolites during soil aquifer treatment. *Water Research*, **37**(15), 3672-3681.

Nancarrow, B., Kaercher, J. and Po, M. (2002). Community Attitudes to Water Restrictions Policies and Alternative Sources: A Longitudinal Analysis 1988–2002. *Australian Research Centre for Water in Society Report*.

Nellor, M. H., Baird, R. B. and Smyth, J. R. (1985). Health effects of indirect potable water reuse. *Journal-American Water Works Association*, **77**(7), 88-96.

Nema, P., Ojha, C., Kumar, A. and Khanna, P. (2001). Techno-economic evaluation of soil-aquifer treatment using primary effluent at Ahmedabad, India. *Water Research*, **35**(9), 2179-2190.

NRC. (1994). Groundwater Recharge Using Waters of Impaired Quality. National Research Council. National Academy Press. Washington D. C., USA.

NRC. (2008). Prospects for Managed Underground Storage of Recoverable Water. National Academy Press, Washington D. C., USA.

NRC. (2012). Water Reuse: Potential for Expanding the Nation's Water Supply Through Reuse of Municipal Wastewater, National Research Council. National Academy Press, Washington, D. C.

Patterson, B. M., Shackleton, M., Furness, A. J., Pearce, J., Descourvieres, C., Linge, K. L., Busetti, F. and Spadek, T. (2010). Fate of nine recycled water trace organic contaminants and metal(loid)s during managed aquifer recharge into a anaerobic aquifer: Column studies. *Water Research*, **44**(5), 1471-1481.

Pescod, M. (1992). Wastewater Treatment and Use in Agriculture. Food and Agriculture Organization Irrigation and Drainage Paper 47. Rome

Po, M., Kaercher, J. and Nancarrow, B. (2003). Literature Review of Factors Influencing Public Perceptions of Water Reuse. Commonwealth Scientific and Institutional Organization (CSIRO), Australia.

Quanrud, D., Arnold, R., Wilson, L. and Conklin, M. (1996). Effect of soil type on water quality improvement during soil aquifer treatment. *Water Science and Technology*, **33**(10), 419-432.

Quanrud, D. M., Carroll, S. M., Gerba, C. P. and Arnold, R. G. (2003). Virus removal during simulated soil-aquifer treatment. *Water Research*, **37**(4), 753-762.

Reed, S. C., Crites, R. and Wallace, A. (1985). Problems with Rapid Infiltration: A Post Mortem Analysis. Journal of Water Pollution Control Federation, 854-858.

Salgot, M. (2008). Water reclamation, recycling and reuse: implementation issues. *Desalination*, **218**(1-3), 190-197.

Sara, M. N. (2003). Site assessment and remediation handbook. 2nd ed. CRC Press.

Sharma, S., Ernst, M., Hein, A., Jekel, M., Jefferson, B. and Amy, G. (2012). Treatment Trains Utilising Natural and Hybrid Processes. *In Water Reclamation Technologies for Safe Managed Aquifer Recharge*, eds. C. Kazner, T. Wintgens and P. Dillon, IWA Publishing. UK, pp. 239-257, ISBN 978-184-339-3443.

Sharma, S., Harun, C. and Amy, G. (2008). Framework for assessment of performance of soil aquifer treatment systems. *Water Science and Technology*, **57**(6), 941-946.

Sharma, S. K., Katukiza, A. and Amy, G. L. (2007). Effect of wastewater quality and process parameters on removal of effluent organic matter (EfOM) during soil aquifer treatment. *Proceedings of ISMAR6 Conference*, Phoenix, USA. 272-284.

Tchobanoglous, G., Burton, F. L. and Stensel, H. D. (2003). *Wastewater Engineering: Treatment and Reuse. 4th ed.* Metcalf and Eddy. McGraw-Hill Company. ISBN 0-07-112250-8.

Toze, S. (1997). Microbial Pathogens in Wastewater. CSIRO Land and Water Technical Report, **1**, 97.

Tredoux, G., Genthe, B., Steyn, M. and Germanis, J. (2012). Managed aquifer recharge for potable reuse in Atlantis, South Africa. *Reclaim Water: Advances in Water Reclaimation Technologies for Safe Managed Aquifer Recharge*, 121.

UNEP. (2003). The Seven "C"s for the Successful Transfer and Uptake of Environmentally Sound Technologies International Environmental Technology Centre, United Nations Environment Programme. Osaka, Japan.

USEPA. (2012). Guidelines for Water Reuse. US Environmental Protection Agency. Office of Research and Development, Cincinnati, Ohio, USA.

USEPA. (1984). Process Design Manual: Land treatment of municipal wastewater. EPA/625/1-81-013a, US Environmental Protection Agency, Cincinnati, Ohio, USA.

USEPA. (1981). Process Design Manual: Land treatment of municipal wastewater EPA /625/1-81-013, US Environmental Protection Agency. Center for Environmental Research Information, Cincinnati, Ohio, USA.

USEPA. (2006). Process design manual: Land Treatment of Municipal Wastewater Effluents. EPA/625/R-06/016, US Environmental Protection Agency, Office of Research and Development, Cincinnati, Ohio, USA.

van Houtte, E., Cauwenberghs, J., Weemaes, M. and Thoeye, C. (2012). Indirect potable reuse via managed aquifer recharge in the Torreele/St-André project. *In Advances in Water Reclamation Technologies for Safe Managed Aquifer Recharge*, IWA Publishing. UK, pp. 33-44.

Vigneswaran, S. and Sundaravadivel, M. (2004). Recycle and Reuse of Domestic Wastewater. *Saravanamuthu (Vigi) Vigneswaran], in Encyclopedia of Life*

Support Systems (EOLSS), Developed under the Auspices of the UNESCO, Eolss Publishers, Oxford UK,[http://www. eolss. net][Retrieved April 24, 2006].

Warner, W. (2000). Influence of religion on wastewater treatment: A consideration for sanitation experts. *Water* **21**, 11-13.

Westerhoff, P. and Pinney, M. (2000). Dissolved organic carbon transformations during laboratory-scale groundwater recharge using lagoon-treated wastewater. *Waste Management,* **20**(1), 75-83.

Zhang, Z., Lei, Z., Zhang, Z., Sugiura, N., Xu, X. and Yin, D. (2007). Organics removal of combined wastewater through shallow soil infiltration treatment: A field and laboratory study. *Journal of Hazardous Materials,* **149**(3), 657-665.

CHAPTER 3

EFFECT OF PRE-TREATMENT OF PRIMARY EFFLUENT USING ALUMINUM SULFATE AND IRON CHLORIDE ON REMOVAL OF SUSPENDED SOLIDS, BULK ORGANIC MATTER, NUTRIENTS AND PATHOGENS INDICATORS[1]

SUMMARY

Aluminium and iron salts are widely used as coagulants in water and wastewater treatment. Two soil columns were fed with coagulated and non-coagulated primary effluent (PE) to investigate the effect of coagulation on removal of suspended solids, bulk organic matter, nitrogen and pathogens indicators during managed aquifer recharge (MAR). Aluminium sulfate and iron chloride were used as coagulants. Experimental results showed considerable suspended solids removal of >65% by both coagulants at optimum doses. However, the overall suspended solids removal by infiltration only and coagulation-infiltration was ~90%. Likewise, removal of phosphorus by coagulation was 80%, whereas total removal by coagulation-infiltration was >98% compared with ~30% by infiltration only. Coagulation of PE removed 16-22% of dissolved organic carbon whereas total removal of ~70% by coagulation-infiltration accounted to 1.4 orders of magnitude higher than infiltration only. Furthermore, removal of pathogens indicators increased considerably from 2.5 $\log_{10}$ units during infiltration only to 3.8 $\log_{10}$ units during coagulation-infiltration for *Escherichia Coli*. Similarly, *total coliforms* removal increased from 2.6 to >4 $\log_{10}$ units. This study showed that coagulation of PE using both aluminium sulfate and iron chloride essentially gives similar removal of the contaminants analyzed. Removal of suspended solids improves operation of SAT site by reducing surface clogging while reduction of phosphorus through coagulation also improved removal of indicator micro-organisms.

---

[1] Based on Abel et al. (2014). *Desalination and Water Treatment, In Press.*

[1] Based on Abel et al. (2013a). Proceedings of IWA Resuse Conference. October 27-31,Windhoek, Namibia

3.1 INTRODUCTION

Wastewater reclamation through managed aquifer recharge (MAR) systems like soil aquifer treatment (SAT) is becoming increasingly attractive in water-scarce regions due to lower capital investment, energy requirements and operator expertise requirements (Sharma et al., 2011; Viswanathan et al., 1999). Nonetheless, wastewater effluents contain high concentrations of suspended solids (SS), colloidal particles and nutrients that interact with soil and cause physical, biological and chemical clogging of the infiltrating surface (Bouwer et al., 2002). Coagulation is a treatment process employed to reduce or neutralize the electric charges on suspended particles or zeta potential (Ebeling et al., 2003) through compression of the electric double layer around colloidal particles leading to formation of microflocs (Matilainen et al., 2010). Coagulation removes colloidal materials which enhance colour and turbidity in wastewater (Amuda et al., 2006). Furthermore, it removes SS and attenuates organic materials to some extent (Al-Mutairi et al., 2004). Flocculation is the agglomeration of the microfloc particles into larger settleable flocs (Amuda and Amoo, 2007; Bratby, 2006). Coagulation/flocculation process allows the van der Waals force of attraction to propel formation of microflocs through aggregation of colloidal and fine suspended materials (Ebeling et al., 2003). Collision between colloidal particles or fine suspended materials is brought about via velocity gradients produced within the fluid by hydraulic or the mechanical means applied for this purpose. However, the coagulation process is influenced by raw water characteristics, temperature, pH, coagulant type, coagulant dose, intensity and duration of rapid mixing.

Aluminium and iron salts are the commonly used inorganic coagulants (Duan and Gregory, 2003). Aluminium coagulants include aluminium sulfate, aluminium chloride, sodium aluminate, aluminium chlorohydrate, polyaluminium chloride, polyaluminium sulfate chloride, polyaluminium silicate chloride. Furthermore, iron coagulants encompass ferric sulfate, ferrous sulfate, chlorinated ferrous sulfate and ferric chloride (Bratby, 2006). While aluminium sulfate (alum, $Al_2(SO_4)_3.18H_2O$) is the most used aluminium salt in water and wastewater treatment (Muyibi et al., 2001), iron chloride ($FeCl_3.6H_2O$) is a widely used iron coagulant (Amuda and Amoo, 2007). Alum coagulation is normally effective within pH range between 6.5 and 7.5. On the other hand, coagulation with iron salts is effective in a pH range between 4 and 11 (Ebeling et al., 2003).

Despite the successful use of aluminium sulfate and iron chloride in water and wastewater treatment processes that apply coagulation, the use of both coagulants for pre-treatment of primary effluent (PE) and subsequent use in SAT has not been well explored.

The purpose of this study was to investigate the use of coagulated PE using aluminium sulfate and iron chloride in SAT with respect to SS, DOC, NH_4-N, PO_4-P and pathogens indicators removal.

3.2   MATERIALS AND METHODS

3.2.1   Source and characteristics of primary effluent

The PE used in this study was sampled from an activated sludge wastewater treatment plant (WWTP) namely, Harnaschpolder, situated in The Hague area, the Netherlands. PE was stored in a cooling room in the laboratory at 4°C upon collection and used within three days after collection. The PE samples were then removed from the cooling room and aerated at room temperature for 4 hours to increase dissolved oxygen (DO) concentration before experimentation. The PE was then allowed to settle after which the supernatant was siphoned and filtered through 63 μm microsieve prior to application to the soil column. Detailed water quality characteristics of the PE used in the study are presented Table 3.1.

Table 3.1 Characteristics of primary effluent

Parameter	Unit	Average concentration
DOC (before aeration)	mg/L	49.4±3.9
DOC (after aeration)	mg/L	41.5±6.8
pH	-	7.4±0.2
Temperature	°C	20.5±0.7
DO (PE upon collection)	mg/L	0.5±0.1
DO (after aeration)	mg/L	7.1±0.8
EC	μs/cm	1407.2±116.3
BOD_5	mg/L	145±7.1
COD	mg/L	448±10.4
SS	mg/L	180±12.7
Turbidity	NTU	139.8±24.2
NO_3-N	mg/L	2.2±0.2
NH_4-N	mg/L	46.5±7.5
PO_4-P	mg/L	10.5±2.3
Alkalinity	mg/L as HCO_3^-	483.1±7.3

3.2.2   Jar test

Analytical grades hydrated iron (ferric) chloride ($FeCl_3.6H_2O$) and aluminium sulfate ($Al_2(SO_4)_3.18H_2O$) from Merck KgaA, Germany were used as the source of iron and aluminium ions utilized for coagulation of PE used in the experiments. A jar test apparatus (VELP SCIENTIFICA JLT6, Italy) with multiple stirrers operated at the same speed was used for rapid and slow mixing. Rapid mixing of PE in 1 L beaker was performed by conducting coagulation at 100 revolutions per minute (rpm) (mean velocity gradient 61.9 s^{-1}) and 150 rpm (mean velocity gradient 113.7 s^{-1}) at 1.5 and 1 minute, respectively. Dosages of iron chloride and aluminium sulfate were added few milliliters below the PE surface at the center of the beakers and varied from 0, 1.93, 3.86, 5.79, 7.72, 9.65, 19.3, 28.95, 38.60, 48.25 and 57.9 mg Fe^{3+}/L for iron chloride and 0, 0.47, 0.93, 1.86, 2.79, 3.72, 4.65, 6.51, 9.30, 11.16, 13.95 and 18.6 mg Al^{3+}/L for aluminium sulfate. Slow mixing was then carried out at 20 rpm for 20 minutes followed by a sedimentation phase of 30 minutes during which the flocs formed in the solution were left to settle. 100 mL sample volume was siphoned from the supernatant of each beaker. Filtered and unfiltered portions from these samples were used to analyze contaminants of interest using the methods detailed in section 3.2.4.

3.2.3 Experimental setup

Laboratory tests were performed on two typical uPVC columns. The columns were roughened from inside (Charles et al., 2008) to minimize preferential and column interface flow, wet-packed in layers by allowing silica sand grains (grain size 0.8-1.25 mm) to settle in de-ionized water while tapping gently on the surface of the column using a mallet rubber hammer to ensure homogeneous media packing in the column. Each column was 4.2 m high with an internal diameter of 57 mm. The media was sieved through 2 mm mesh screen followed by cone and splitting to obtain a representative sample as detailed in Schumacher et al. (1991). A ponding headspace of 20 cm was provided on the top of each column and influent samples were taken from a port situated in this length to account for any PE quality change in the connection tubes between the feed tank and the column headspace. Sampling ports (SPs) were fitted at an interval of 10 cm in the upper 50 cm of the medium after which subsequent sampling ports were deployed at 50 cm intervals. The ports were extended to the middle of the column using glass tubes with 5 mm diameter. The bottom 20 cm of the column was filled with gravel with grain size ranging from 2 mm to 10 mm as a support layer. A variable-speed peristaltic pump was used to continuously deliver the PE to the top of the columns at hydraulic loading rate (HLR) of 1.25 m/d at room temperature. This HLR was frequently checked at inlet and outlet points of each column using a measuring cylinder and a stopwatch. Hydraulic residence time (HRT) of 3.2 days was obtained under continuous PE loading (wetting) at this HLR.

Biofilms formation on the media and subsequent bio-stability of the soil columns was monitored for 120 days by analyzing dissolved organic carbon (DOC) of influent and effluent samples filtered through 25 mm diameter regenerated cellulose filter with nominal size of 0.45 μm (Whatman, Germany) after which DOC removal was calculated. The columns were assumed to be ripened (bio-stablized) when a difference of ±1 % DOC removal was obtained between each successive pair of samples. Ripening process was repeatedly carried out when operating conditions of the system were changed to ensure that the microorganisms have adapted to the new environmental conditions. Samples from the ripened columns (PE + SAT$_{Al}$ and PE + SAT$_{Fe}$) were analyzed for suspended solids (SS), DOC, nitrogen, phosphorus and pathogens indicators. PE coagulated with aluminium sulfate and iron chloride was then infiltrated into columns. While the first column received PE coagulated with aluminium sulfate (PE + COAG Al³⁺ + SAT), the second column was fed with PE pre-treated using iron chloride (PE + COAG Fe³⁺ + SAT).

During infiltration of both coagulated and non-coagulated PE, the infiltration rate decreased due to formation of a clogging layer at the surface of the media which was frequently flushed with new PE gently added to the inner surface of the column and removed by syringe. This process was performed once the infiltration rate drops by 50%. Figure 3.1 illustrates typical schematic of the soil column used in these experiments.

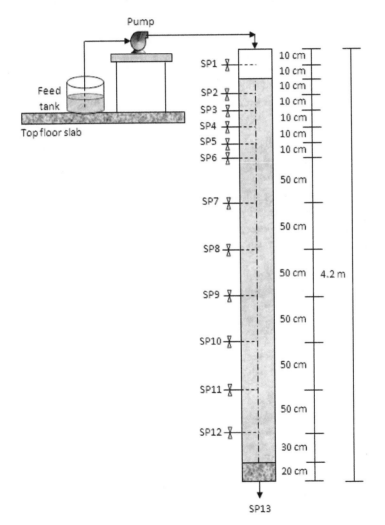

Figure 3.1 Schematic of soil column experimental setup

3.2.4 Analytical methods

DOC concentrations of all pre-filtered samples collected from the WWTP and laboratory-based set-ups were determined within three days by the combustion technique using total organic carbon analyzer (TOC-VCPN (TN), Shimadzu, Japan) with a precision range of 0.5 to 20 mg/L. Milli-Q water (Advantage A10, Millipore) and control samples with known DOC concentrations were analyzed alongside the samples to ensure that the TOC analyzer gives reliable measurements. Since aromatic unsaturated bonding structures in organic compounds are responsible for ultraviolet (UV) light absorption over the range of 200 - 300 nm (Michail and Idelovitch, 1981), UV absorbance at wavelength of 254 nm (UVA_{254}) for all pre-filtered samples was measured in a quartz cell (cuvette) with a 1 cm path length using a computer controlled UV-VIS spectrophotometer (UV-2501 PC, Shimadzu, Japan). The instrument was auto-zeroed prior to sample measurements to obtain zero absorbance (reference) using ultra-pure water (Milli-Q). Specific ultraviolet light absorbance (SUVA) was calculated to explore the contribution of aromatic structures of DOC of the samples using their UVA_{254} measurements and corresponding DOC values (UVA_{254} x 100/DOC).

Chemical reagents used to determine ammonium as nitrogen (NH_4-N), nitrate as nitrogen (NO_3-N) and phosphate as phosphorus (PO_4-P) were of analytical grade and were purchased from Merck KGaA, Germany and J.T. BAKER, Netherlands. NH_4-N, NO_3-N and PO_4-P were determined using colorimetric automated techniques using a spectrophotometer according to Eaton et al. (2005). Standard calibration line (in 5 concentration range) was prepared for NH_4-N, NO_3-N and PO_4-P to calculate their concentrations in various water samples. Determination of SS was performed by drying a 47 mm diameter regenerated cellulose filter with nominal size of 0.4-1 μm (Whatman, Germany) in a furnace at 520°C for 3 h. The filter was weighed before and after filtering a well-mixed 50 mL of PE and dried at 105°C for at least 2 h until a constant weight was obtained. SS was then calculated as the difference in the filter weight relative to sample volume used.

Plate count method was used to enumerate *Escherichia Coli* (*E. coli*) and *total coliforms* in unfiltered water samples from soil columns. 26.5 g of chromocult agar (Merck KGaA, Germany) was dissolved in 1 L of deionized water in a round-bottom flask and placed in a water bath at a temperature of 99°C for 30 minutes. The flask was then removed from the water bath and kept at a temperature of 50°C in an oven for 30 minutes after which it was decanted into two smaller (0.5 L each) round-bottom flasks. Finally, the warm liquid agar was poured into test plates and left to solidify. The plates were placed in refrigerator at 4°C for one week prior experimentation. 0.1 mL from well centrifuged primary effluent was transferred (using pipette) to test plates (in triplicate) and cultured in test plates containing chromocult agar for 24 h at 37°C.

3.3   RESULTS AND DISCUSSION

3.3.1   Coagulation procedure and optimization

During the preliminary experiments, coagulation of PE was performed at different rapid mixing intensities and time to determine the appropriate mixing intensity, time and optimum coagulant concentrations (OCCs) for aluminium sulfate and iron chloride. OCCs for both iron chloride and aluminium sulfate were the concentrations beyond which no further substantial reduction of residual turbidity and SS was achieved. As presented in Figure 3.2 and Table 3.2, no further appreciable reduction of turbidity and SS was achieved beyond aluminium sulfate dosage of 9.3 mg Al^{3+}/L for both 100 and 150 rpm mixing intensities. Removals of 91% turbidity and 77.5% SS were achieved at mixing intensity of 150 rpm compared to 83.3 and 64.1% at 100 rpm. However, PE coagulation using iron chloride and mixing intensities of 100 and 150 rpm revealed no major change in turbidity removal beyond 88.3 and 86.3% at coagulants doses of 28.95 and 19.3 mg Fe^{3+}/L. SS removals of 74.1 and 71.6% were attained under similar coagulant doses and mixing intensities.

Among the two rapid mixing intensities examined, 150 rpm demonstrated higher turbidity and SS removal compared to 100 rpm. It was also noticed that no substantial turbidity and SS removal observed beyond coagulant doses higher than 9.3 mg Al^{3+}/L for aluminium sulfate and 19.3 mg Fe^{3+}/L for ferric chloride. Thus, it was concluded that rapid mixing intensity of 150 rpm and OCCs of 9.3 mg Al^{3+}/L and 19.3 mg Fe^{3+}/L should be adopted for coagulation of the PE used prior to infiltration.

(a) (b)

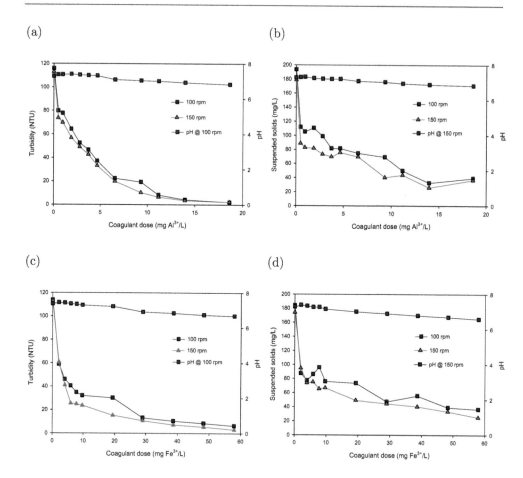

(c) (d)

Figure 3.2 Effect of rapid mixing of PE in 1 L beaker on residual turbidity and suspended solids using aluminium sulfate (a and b) and ferric chloride (c and d)

Table 3.2 Effect of coagulant type, coagulant dose and rapid mixing intensities on removal of turbidity and suspended solids from PE.

Coagulant	Mixing intensity (rpm)	Coagulant dosage (mg/L)	Turbidity removal (%)	Suspended solids removal (%)	pH
Aluminium sulfate	100	9.3	83.3	64.1	7.05
	150	9.3	91	77.5	7.05
Iron chloride	100	28.95	88.3	74.1	7.01
	150	19.3	86.3	71.6	7.12

3.3.2   Suspended solides

Removal of SS is an essential feature of wastewater coagulation (Al-Mutairi et al., 2004). SS was monitored by tracking the quality of infiltrated coagulated and non-coagulated PE along the soil column. It was observed that most SS removal from non-coagulated PE (influent: 170.3 ± 3.7 mg/L) occurred in the upmost part of the column. Coagulation of PE (influent: 185.0 ± 4.2 mg/L) using OCC of 9.3 mg Al^{3+}/L from aluminium sulfate and OCC of 19.3 mg Fe^{3+}/L from iron chloride removed 65.1 ± 0.4 and $68.7\pm0.8\%$, respectively. However, similar total SS removals of 90.4 ± 0.7, 89.6 ± 0.7, 89.9 ± 0.5 and $90.9\pm0.1\%$ were achieved in PE + SAT_{Al}, PE + SAT_{Fe}, PE + COAG Al^{3+} + SAT and PE + COAG Fe^{3+} + SAT, respectively. These results suggest that despite of substantial initial SS removal by coagulation process, there was no notable contribution to the total SS removal as compared to infiltration only. Most of SS was removed in the upmost 1 m of the column during infiltration of non-coagulated PE. However, the coagulation process led to notable reduction of clogging layer development at the surface of the media (data not provided). This process prolonged the frequency at which the media surface was gently flushed with new PE from one week (during infiltration of non-coagulated PE) to four weeks. Furthermore, equal SS removals following PE coagulation with both coagulants implies that the cost of each coagulant coupled with the volume of sludge generated would dominate which coagulant should be adopted for pre-treatment of PE prior to infiltration.

3.3.3   Bulk organic matter

Aluminium sulfate and iron chloride act to destabilize and remove colloidal and dissolved organic carbon through production of cationic hydrolysis products (Sharp et al., 2006). For both coagulants, DOC content of 44.4 ± 3.4 mg/L was removed at optimum coagulant doses by $16.3\pm1.1\%$ with aluminium sulfate and $22.2\pm0.5\%$ with iron chloride. These observed differences are consistent with the findings of other researchers that higher DOC removals are achieved with iron chloride as compared to aluminium sulfate (Bell-Ajy et al., 2000; Edzwald and Tobiason, 1999) which is attributable to formation of bigger and stronger flocs by iron chloride (Ratnaweera et al., 1999; Stephenson and Duff, 1996) that facilitate DOC charge neutralization, adsorption and entrapment into insoluble aggregates (Jarvis et al., 2005). Furthermore, when the coagulated PE was infiltrated into soil columns total DOC removals of 68.1 ± 2.5 and $70.4\pm0.3\%$ were attained in PE + COAG Al^{3+} + SAT and PE + COAG Fe^{3+} + SAT, respectively. These results show no considerable difference in using either of the coagulants to pre-treat PE. However, DOC removal was relatively higher than that achieved during infiltration of non-coagulated PE where 47.6 ± 1.5 and $49.3\pm0.4\%$ of initial DOC concentration (35.7 ± 1.0 mg/L) was removed in PE + SAT_{Al} and PE + SAT_{Fe}, respectively. These results show that wastewater coagulation prior to biological treatment (i.e. infiltration) enhances DOC biodegradability during biological treatment (Amuda and Alade, 2006). Figure 3.3 presents DOC concentration profiles along soil columns fed with PE.

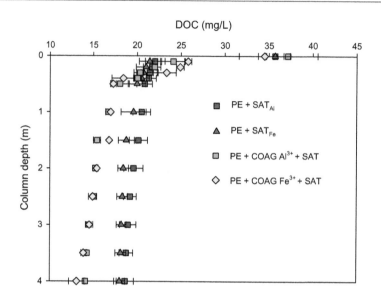

Figure 3.3 Average DOC concentration as a function of column depth fed with PE (media size: 0.8–1.25 mm, HLR = 0.625 m/d)

SUVA values changed during infiltration of both coagulated and non-coagulated PE. Coagulation process using aluminium sulfate and iron chloride reduced SUVA of PE slightly from 1.9±0.1 L/mg. m to respective values of 1.7±0.1 and 1.5±0.1 L/mg. m suggesting removal of aromatic fractions of DOC. On the contrary, SUVA values of the coagulated PE upon infiltration increased considerably in the first 1 m of the column followed by decrease in the subsequent 3 m. SUVA values along PE + COAG Al^{3+} + SAT profile increased from an influent value of 1.8±0.1 L/mg. m to 3.6±0.2 L/mg. m in the first 1 m of the column, then slightly decreased to 2.4±0.1 L/mg. m at the effluent. Similar trends were observed along PE + COAG Fe^{3+} + SAT profile where SUVA increased from 1.5±0.1 to 3.4±0.1 L/mg. m in the upmost 1 m before decreased to 2.5±0.2 L/mg. m at the effluent. On the other hand, SUVA of the non-coagulated PE increased from an influent value of 1.7±0.1 to 3.0±0.2 L/mg. m in the upmost 1 m of the column after which a steady decrease was observed along the remaining 3 m depth of the column, resulting in an effluent SUVA value of 2.3±0.0 L/mg. m. Increase in SUVA values in the first 1 m of the column is attributed to removal of aliphatic DOC fractions, whereas decrease in SUVA values suggests removal of aromatic DOC fractions in the subsequent 3 m of the column.

3.3.4 Nitrogen

Coagulation of PE prior to infiltration resulted in initial NH_4-N reduction of 6.6±1.3 and 7.2±0.2% by aluminium sulfate and iron chloride, respectively. However, coagulation and infiltration combined led to a total NH_4-N reduction of 60.9±2.6 and 64.7±1.9% in PE + COAG Al^{3+} + SAT and PE + COAG Fe^{3+} + SAT, respectively.

Infiltration of non-coagulated PE removed 21.4±0.9 and 22.9±2.4% of its NH_4-N content of 48.9±3.9 mg N/L in in PE + SAT_{Al} and PE + SAT_{Fe}, respectively. It was noticed that most of NH_4-N reduction occurred in the first 1 m through nitrification process as evidenced by the corresponding increase in NO_3-N concentrations at this depth. Further NH_4-N reduction observed along the column was attributed to adsorption of NH_4-N to the media (Paranychianakis et al., 2006). On the other hand, NO_3-N content of non-coagulated PE were 1.0±0.2, 5.4±0.3 and 2.4±0.3 mg N/L at influent, <1 m and 4 m respectively. At the same depths, coagulated PE exhibited 1.2±0.1, 14.2±1.6 and 7.2±0.7 mg N/L NO_3-N concentrations. A steady decrease of NO_3-N was observed in the last 3 m of the column suggesting dominance of denitrification process as evidenced by low NO_3-N concentration exiting the column. Removal of NO_3-N by denitrification process in MAR systems prevents pollution of groundwater (Paranychianakis et al., 2006). DO concentration rapidly decreased from 7.8±0.3 mg/L to as low as 0.6±0.1 mg/L in the first 1 m of the column suggesting utilization of DO by microorganisms to mediate biological reduction of DOC and NH_4-N. Figure 3.4 illustrates change in (a) NH_4-N concentrations along columns profiles and (b) corresponding changes in NO_3-N concentrations.

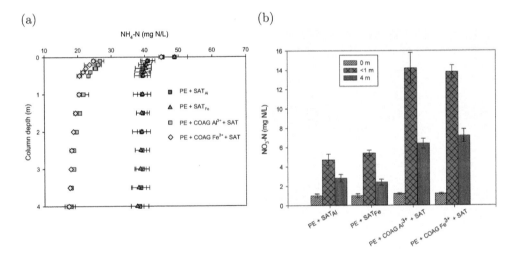

Figure 3.4 Average (a) NH_4-N and (b) NO_3-N concentrations profiles along the depth of soil column fed with PE and operated at alternate wetting/drying cycles (media size: 0.8–1.25 mm, HLR = 0.625 m/d)

3.3.5 Phosphorus

Substantial removal of PO_4-P from PE was achieved through coagulation. Aluminium sulfate removed 79.1±1.2% of influent PO_4-P of 12.6±0.1 mg/L, whereas iron chloride removed 80.3±0.5%. Infiltration of the coagulated PE diminished the remaining PO_4-P content leading to a total removal of 98.4±0.2 and 98.1±0.3% in PE + COAG Al^{3+} + SAT and PE + COAG Fe^{3+} + SAT, respectively. However, infiltration of non-coagulated PE removed 30.6±0.7% of influent PO_4-P of 10.1±0.6 mg/L, whereas iron chloride removed $29.4\pm2.0\%$. PO_4-P is mainly sorbed or precipitated in filter media (Vohla et al., 2007). The main removal mechanism for PO_4-P during infiltration is predominantly adsorption, which diminishes once the sorption capacity of the media is exhausted due to continuous application of PO_4-P (Paranychianakis et al., 2006). This limitation has presumably caused the relative low PO_4-P removal from the non-coagulated PE. Furthermore, the presence of natural organic matter (NOM) originating from surface water used as drinking water source led to competition between the NOM and PO_4-P for adsorption sites, especially at pH above 7 (7.4 ± 0.2) (El Samrani et al., 2004). Figure 3.5 shows change in PO_4-P concentrations of coagulated and non-coagulated PE along soil columns profiles.

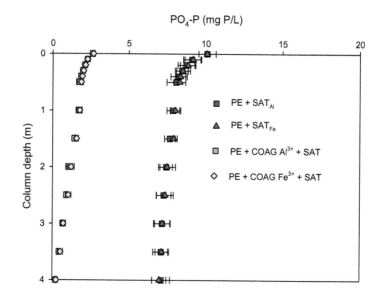

Figure 3.5 Average PO_4-P concentration along the depth of soil column fed with PE (media size: 0.8–1.25 mm, HLR: 0.625 m/d)

3.3.6 *E. Coli* and *total coliforms* removal

The effect of coagulation followed by infiltration on removal of *E. coli* and *total coliforms* from PE was examined using two coagulants. The mean concentrations of indicator pathogens in PE treated with coagulation was $6.6\times10^6\pm0.6\times10^6$ CFU/100 mL for *E. coli* and $25.1\times10^6\pm0.9\times10^6$ CFU/100 mL for *total coliforms*. Coagulation of PE using aluminium sulfate removed 0.6 ± 0.2 $\log_{10}$ units of *E. coli* compared to 0.6 ± 0.1 $\log_{10}$ units of *total coliforms*. However, coagulation using iron chloride removed 0.9 ± 0.3 and 0.6 ± 0.0 $\log_{10}$ units of *E. coli* and *total coliforms*, respectively. From Table 3.3, it can be seen that infiltration of non-coagulated PE removed 2.5 ± 0.2 $\log_{10}$ units of *E. coli* and 2.7 ± 0.3 $\log_{10}$ units of *total coliforms* in PE + SAT_{Al}

compared to 2.5 ± 0.1 $\log_{10}$ units of *E. coli* and 2.6 ± 0.3 $\log_{10}$ units of *total coliforms* in $PE + SAT_{Fe}$.

Table 3.3 Effect of PE coagulation and infiltration in soil columns (influent: primary effluent, media size: 0.8–1.25 mm, HLR: 0.625 m/d)

Feed water and operating condition	Pathogens indicator removal ($\log_{10}$ units)	
	E. coli	*total coliforms*
$PE + SAT_{Al}$	2.5 ± 0.2	2.7 ± 0.3
$PE + SAT_{Fe}$	2.5 ± 0.1	2.6 ± 0.3
$PE + COAG\ Al^{3+} + SAT$	3.8 ± 0.0	4.4 ± 0.0
$PE + COAG\ Fe^{3+} + SAT$	3.8 ± 0.0	4.3 ± 0.0

E. coli and *total coliforms* removal in $PE + COAG\ Al^{3+} + SAT$ accounted to 3.8 ± 0.0 and 4.4 ± 0.0 $\log_{10}$ units, respectively. Similar removal was achieved when PE was coagulated with iron chloride then infiltrated into $PE + COAG\ Fe^{3+} + SAT$ resulting in *E. coli* and *total coliforms* removal of 3.8 ± 0.0 and 4.3 ± 0.0 $\log_{10}$ units, respectively. Attenuation of pathogens during infiltration is achieved through inactivation, straining and attachment to aquifer materials (McDowell-Boyer et al., 1986). High removals for both *E. coli* and *total coliforms* by coagulation coupled with infiltration might be attributed to removal of PO_4-P by coagulation which reduced competition between pathogens indicators and PO_4-P for adsorption sites in the media along the column profile.

PE coagulation prior to infiltration is an attractive option for SAT as it reduces SS content of PE substantially and consequently reduces the frequency at which the clogging layer develops. Besides, introduction of coagulation improved the overall DOC, NH_4-N and pathogens removal regardless of the coagulant type. A summary of contaminant removal is presented in Table 3.4.

Table 3.4 Summary of contaminants removal during infiltration of coagulated and non-coagulated primary effluent in soil columns (media size: 0.8–1.25 mm, HLR: 0.625 m/d)

Operating conditions	Contaminant removal					
	SS (%)	DOC (%)	NH_4-N (%)	PO_4-P (%)	*E. coli* ($\log_{10}$ units)	*Total coliforms* ($\log_{10}$ units)
$PE + SAT_{Al}$	90.4 ± 0.7	47.6 ± 1.5	21.4 ± 0.9	30.6 ± 0.7	2.5 ± 0.2	2.7 ± 0.3
$PE + SAT_{Fe}$	89.6 ± 0.7	49.3 ± 0.4	22.9 ± 2.4	29.4 ± 2.0	2.5 ± 0.1	2.6 ± 0.3
$(PE + COAG\ Al^{3+} + SAT)$	89.9 ± 0.5	68.1 ± 2.5	60.9 ± 2.6	98.4 ± 0.2	3.8 ± 0.0	4.4 ± 0.0
$(PE + COAG\ Fe^{3+} + SAT)$	90.9 ± 0.1	70.4 ± 0.3	64.7 ± 1.9	98.1 ± 0.3	3.8 ± 0.0	4.3 ± 0.0

It is clear from this study that coagulation of PE is a viable pre-treatment for MAR systems as the water quality is substantially improved and the clogging potential is reduced. In areas where land costs are relatively high, coagulation of PE will help to reduce surface clogging and consequently the area required for a SAT basin as relatively higher infiltration rates can be applied with coagulated PE. However, the capital and O&M costs (including sludge management) of the coagulation of PE should be compared with other alternatives for pre-treatment at local level, before selecting the most appropriate pre-treatment option (Sharma et al., 2011).

Aluminium sulfate could be used to treat water with high alkalinity since it reacts with natural alkalinity in water and causes the pH reduction and limits coagulation efficiency to pH range 6.5 and 7.5 (Ebeling et al., 2003; Ndabigengesere and Subba Narasiah, 1998). To maintain this pH range, addition of considerable amount of lime to the water is normally required (Ebeling et al., 2003). Such practice produces a large quantity of sludge that requires disposal and questions the assumption that aluminium sulfate is cheaper than iron chloride. Iron chloride suites water with both low and high natural alkalinity and remains efficient at a wider pH range of 4 and 11 (Ebeling et al., 2003). Furthermore, coagulation with iron chloride produces numerous, stronger, and heavier flocs compared to those formed during coagulation with aluminium sulfate (Ratnaweera et al., 1999; Stephenson and Duff, 1996). It is therefore recommended to use the most affordable coagulant which meets pre-treatment water quality and economic requirements taking into account sludge management and size of treatment units required.

Where metal based (inorganic) coagulants are not affordable for water and wastewater treatment in some developing countries (Ndabigengesere and Subba Narasiah, 1998), the use of locally available natural (organic) coagulants such as chitosan, alginates (Bratby, 2006), okra (Diaz et al., 1999) and *moringa oleifera* (Ndabigengesere and Subba Narasiah, 1998) becomes indispensible. However, these coagulants should be individually checked for adverse impacts associated with their use since some of these natural (organic) coagulants could enhance the organic matter content of the water and consequently the potential of disinfection byproducts (DBPs) formation if the reclaimed water is treated with chlorine (Cl_2).

3.4 CONCLUSIONS

Coagulation process using aluminium sulfate and iron chloride reduced the initial SS load of PE prior to application to the soil column by >65%. This considerable SS removal has a positive effect on SAT site performance in terms of infiltrative surface clogging. Nevertheless, there was no major effect on the overall SS removal following infiltration of the coagulated PE which levelled at ~90%, similar to SS removal obtained from infiltration of non-coagulated PE.

Coagulation of PE using aluminium sulfate and iron chloride resulted in DOC removal of 16.3±1.1 and 22.2±0.5%, respectively. Infiltration of the coagulation PE led to a profound DOC removal which was 1.4 orders of magnitude higher than infiltration only for both coagulants. Substantial PO_4-P removal (>80%) was achieved by coagulation of PE using aluminium sulfate and iron chloride. Furthermore, infiltration of non-coagulated PE removed ~30% of PO_4-P compared to >98% removal by coagulation and infiltration combined.

Removal of *E. coli* and *total coliforms* by coagulation was less than 1 log_{10} units. However, infiltration of coagulated PE led to 3.8 and >4 log_{10} units for both pathogens indicators compared to ~2.5 log_{10} units for non-coagulated PE. This ascribed to the fact that PO_4-P competes with pathogens indicators for adsorption sites in non-coagulated PE. This competition was substantially reduced by removal of PO_4-P through coagulation which in turn led to relatively high bacteria removal.

In general, it can be concluded that coagulation of PE is an attractive pre-treatment alternative to improve the overall performance of SAT using PE. Both coagulants can be equally used to improve operation and maintenance of SAT system, reduce land area and minimizes post-treatment requirements for reclaimed water. Nevertheless, the use of organic coagulants (i.e. *moringa oleifera*, chitosan and okra) should be explored for PE pre-treatment where inorganic coagulants are not affordable.

3.5 REFERENCES

Al-Mutairi, N., Hamoda, M. and Al-Ghusain, I. (2004). Coagulant selection and sludge conditioning in a slaughterhouse wastewater treatment plant. *Bioresource Technology,* **95**(2), 115-119.

Amuda, O., Amoo, I. and Ajayi, O. (2006). Performance optimization of coagulant/flocculant in the treatment of wastewater from a beverage industry. *Journal of Hazardous Materials,* **129**(1), 69-72.

Amuda, O. S. and Alade, A. (2006). Coagulation/flocculation process in the treatment of abattoir wastewater. *Desalination,* **196**(1-3), 22-31.

Amuda, O. S. and Amoo, I. A. (2007). Coagulation/flocculation process and sludge conditioning in beverage industrial wastewater treatment. *Journal of Hazardous Materials,* **141**(3), 778-783.

Bell-Ajy, K., Abbaszadegan, M., Ibrahim, E., Verges, D. and LeChevallier, M. (2000). Conventional and optimized coagulation for NOM removal. *Journal of American Water Works Association,* **92**(10), 44-58.

Bouwer, H., Rice, R., Lance, J. and Gilbert, R. (2002). Artificial recharge of groundwater: hydrogeology and engineering. *Hydrogeology Journal,* **10**(1), 121-142.

Bratby, J. (2006). *Coagulation and Flocculation in Water and Wastewater Treatment.* 2nd ed. IWA Publishing, London; Seattle.

Charles, K. J., Souter, F. C., Baker, D. L., Davies, C. M., Schijven, J. F., Roser, D. J., Deere, D. A., Priscott, P. K. and Ashbolt, N. J. (2008). Fate and transport of viruses during sewage treatment in a mound system. *Water Research,* **42**(12), 3047-3056.

Diaz, A., Rincon, N., Escorihuela, A., Fernandez, N., Chacin, E. and Forster, C. (1999). A preliminary evaluation of turbidity removal by natural coagulants indigenous to Venezuela. *Process Biochemistry,* **35**(3), 391-395.

Duan, J. and Gregory, J. (2003). Coagulation by hydrolysing metal salts. *Advances in Colloid and Interface Science,* **100-102**(0), 475-502.

Eaton, A. D., Clesceri, L. S., Rice, E. W. and Greenberg, A. E. (2005). *Standard Methods for the Examination of Water and Wastewater. 21th ed.* American Public Health Association, American Water Works Association, and Water Environment Federation, Washington, DC., USA.

Ebeling, J. M., Sibrell, P. L., Ogden, S. R. and Summerfelt, S. T. (2003). Evaluation of chemical coagulation-flocculation aids for the removal of suspended solids and phosphorus from intensive recirculating aquaculture effluent discharge. *Aquacultural Engineering,* **29**(1-2), 23-42.

Edzwald, J. K. and Tobiason, J. E. (1999). Enhanced coagulation: US requirements and a broader view. *Water Science and Technology,* **40**(9), 63-70.

El Samrani, A., Lartiges, B., Montargès-Pelletier, E., Kazpard, V., Barres, O. and Ghanbaja, J. (2004). Clarification of municipal sewage with ferric chloride: the nature of coagulant species. *Water Research,* **38**(3), 756-768.

Jarvis, P., Jefferson, B. and Parsons, S. A. (2005). How the natural organic matter to coagulant ratio impacts on floc structural properties. *Environmental Science and Technology,* **39**(22), 8919-8924.

Matilainen, A., Vepsäläinen, M. and Sillanpää, M. (2010). Natural organic matter removal by coagulation during drinking water treatment: A review. *Advances in Colloid and Interface Science,* **159**(2), 189-197.

McDowell-Boyer, L., Hunt, J. and Sitar, N. (1986). Particle transport through porous media. *Water Resources Research,* **22**(13), 1901-1921.

Michail, J. and Idelovitch, E. (1981). Gross organic measurements for monitoring of wastewater treatment and reuse. *Chemistry in Water Reuse,* **1**, 35-64.

Muyibi, S., Megat, J., Hazalizah, H. and Irmayanti, I. (2001). Coagulation of river water with Moringa oleifera seeds and alum–a comparative study. *Journal Institute of Engineers Malaysia,* **62**(2), 15-21.

Ndabigengesere, A. and Subba Narasiah, K. (1998). Quality of water treated by coagulation using Moringa oleifera seeds. *Water Research,* **32**(3), 781-791.

Paranychianakis, N. V., Angelakis, A. N., Leverenz, H. and Tchobanoglous, G. (2006). Treatment of wastewater with slow rate systems: a review of treatment processes and plant functions. *Critical Reviews in Environmental Science and Technology,* **36**(3), 187-259.

Ratnaweera, H., Hiller, N. and Bunse, U. (1999). Comparison of the coagulation behavior of different Norwegian aquatic NOM sources. *Environment International,* **25**(2-3), 347-355.

Schumacher, B., Shines, K., Burton, J. and Papp, M. (1991). A Comparison of soil Sample Homogenization Techniques. In Hazardous Waste Measurements." MS Simmons, ed. Chelsea, Michigan: Lewis Publishers, pp. 53-68.

Sharma, S. K., Hussen, M. and Amy, G. L. (2011). Soil aquifer treatment using advanced primary effluent. *Water Science and Technology,* **64**(3), 640-646.

Sharp, E. L., Parsons, S. A. and Jefferson, B. (2006). The impact of seasonal variations in DOC arising from a moorland peat catchment on coagulation with iron and aluminium salts. *Environmental Pollution,* **140**(3), 436-443.

Stephenson, R. J. and Duff, S. J. B. (1996). Coagulation and precipitation of a mechanical pulping effluent—I. Removal of carbon, colour and turbidity. *Water Research,* **30**(4), 781-792.

Viswanathan, M., Al Senafy, M., Rashid, T., Al-Awadi, E. and Al-Fahad, K. (1999). Improvement of tertiary wastewater quality by soil aquifer treatment. *Water Science and Technology,* **40**(7), 159-163.

Vohla, C., Alas, R., Nurk, K., Baatz, S. and Mander, Ü. (2007). Dynamics of phosphorus, nitrogen and carbon removal in a horizontal subsurface flow constructed wetland. *Science of the Total Environment,* **380**(1-3), 66-74.

CHAPTER 4

IMPACT OF HYDRAULIC LOADING RATE AND SOIL TYPE ON REMOVAL OF BULK ORGANIC MATTER AND NITROGEN FROM PRIMARY EFFLUENT IN LABORATORY-SCALE SOIL AQUIFER TREATMENT SYSTEM[2]

SUMMARY

The effect of hydraulic loading rate (HLR) and media type on the removal of bulk organic matter and nitrogen from primary effluent during soil aquifer treatment was investigated by conducting laboratory-scale soil column studies. Two soil columns packed with silica sand were operated at HLRs of 0.625 m/d and 1.25 m/d, while a third column was packed with dune sand and operated at HLR of 1.25 m/d. Bulk organic matter was effectively removed by $47.5\pm1.2\%$ and $45.1\pm1.2\%$ in silica sand columns operated at 0.625 m/d and 1.25 m/d, respectively and $57.3\pm7.6\%$ in dune sand column operated at 1.25 m/d. Ammonium-nitrogen (NH_4-N) reduction of $74.5\pm18.0\%$ was achieved at HLR 0.625 m/d compared to $39.1\pm4.3\%$ at 1.25 m/d in silica sand columns whereas $49.2\pm5.2\%$ NH_4-N reduction was attained at 1.25 m/d in the dune sand column. NH_4-N reduction in the first 3 m was assumed to be dominated by nitrification process evidenced by corresponding increase in nitrate. Part of the NH_4-N was adsorbed onto the media which was observed at higher rates between 3 and 5 m in silica sand column operated at HLR of 0.625 m/d and dune sand column operated at 1.25 m/d compared to 1.25 m/d silica.

---

[2] Based on Abel et al. (2013). *Water Science and Technology*, **68**(1), 217-226.

4.1 INTRODUCTION

Soil aquifer treatment (SAT) has been employed to provide additional treatment of primary, secondary and tertiary effluents from wastewater treatment plants for reuse purposes (Crites et al., 2006; Fox et al., 2001; Nema et al., 2001; Sharma et al., 2011; Wilson et al., 1995). Primary effluent (PE) is characterized with high ammonium, low nitrate, and relatively high phosphorus concentrations (Ho et al., 1992). SAT offers a wide range of benefits over the conventional wastewater treatment methods such as provision of added storage capacity and low cost (Bouwer, 1991). Results from different SAT sites in many countries have revealed its efficiency to remove organics, nutrients, bacteria, and virus from primary and secondary effluents (Nema et al., 2001). Organic carbon is a major water quality concern in SAT systems that involve indirect potable reuse of the reclaimed water. Organic matter in wastewater effluents is found in the form of effluent organic matter (EfOM) which constitutes natural organic matter (NOM) from drinking water, anthropogenic organic compounds emanating from the domestic use of water, and soluble microbial products (SMPs) produced during wastewater treatment process (Drewes et al., 2006). The type and bioavailability of EfOM affect the extent of soil biomass growth in managed aquifer recharge (MAR) systems (Rauch-Williams et al., 2010; Rauch and Drewes, 2004) such as SAT. The quality of the wastewater applied to MAR is primarily improved in the upper part of the vadose zone (Yamaguchi et al., 1996). Large numbers of bacteria are found at the water-soil interface during application of wastewater effluent to porous media (Emerick et al., 1997) due to presence of high concentrations of organic carbon and nutrients. However, variation in electron acceptors and donors' concentrations during application of wastewater effluent in a MAR system may change the dynamics of microbial community (Dillon et al., 2009a) in the biofilms environment.

The performance of a SAT system is influenced by several factors including (but not limited to) hydraulic loading rate (HLR), soil type, soil organic content and reduction-oxidation (redox) conditions in the soil matrix. Not all soils are appropriate for pollutants removal during SAT (Ho et al., 1992) and suitability of hydogeological properties of aquifers for MAR should be examined (Dillon et al., 2009b). Soils with substantial clay fractions should be avoided during SAT site selection due to their relative impermeability that leads to high land requirements for percolation ponds. On the other hand, coarse sands yield high infiltration capacity, but remove less pollutants (Ho et al., 1992). Hence, fine sand, loamy sand, and sandy loam range are the most suitable soils for SAT (Pescod, 1992).

This chapter reports the results of a laboratory-scale columns studies using primary effluent (PE) and operated at two different HLRs to investigate the effect of HLR on the reduction of bulk organic matter and nitrogen. Besides, the impact of media type on the reduction of bulk organic matter and nitrogen was probed using two different media in laboratory-scale soil columns operated at the same HLR.

4.2 MATERIALS AND METHODS

4.2.1 Source water characteristics

Primary effluent (PE) was collected every week from an activated sludge wastewater treatment plant (WWTP) of Nieuwewaterweg located at Hoek van Holland, The Netherlands. The WWTP receives >90% domestic wastewater combined with pre-treated wastewater from glass houses and industrial wastewater. The PE was thoroughly mixed and aerated until dissolved oxygen (DO) of 6.0 ± 0.5 mg O_2/L was reached prior to application to laboratory-scale columns. Characteristics of PE used are shown in Table 4.1.

Table 4.1 Average water quality characteristics of the PE used

Parameter	Units	Measured value
Temperature	^{0}C	13.2 ± 0.4
Dissolved oxygen	mg O_2/L	2.1 ± 0.5
EC	μS/cm	1497 ± 121.2
pH	-	7.3 ± 0.1
DOC	mg/L	42.5 ± 10.3
UVA$_{254}$	1/cm	1.3 ± 0.2
SUVA	L/mg. m	3.0 ± 0.4
PO$_4$-P	mg/L	5.3 ± 1.7
NH$_4$-N	mg/L	32.0 ± 4.3
NO$_3$-N	mg/L	0.7 ± 0.3
COD	mg/L	296 ± 46
BOD$_5$	mg/L	180.7 ± 34.7
E. coli	CFU/100 mL	$3.6\times10^6\pm1.6\times10^6$
Total coliforms	CFU/100 mL	$3.1\times10^7\pm1.0\times10^7$
Alkalinity	mg/L as HCO$_3^-$	497.0 ± 46.6
Suspended solids (SS)	mg/L	107.1 ± 30.4

4.2.2 Filter media characteristics

Silica sand used in this study was obtained from Filcon (Papendrecht, Netherlands) while dune sand was brought from dune area in The Hague, Netherlands. Homogeneity of both media was assured by sieving through a 2 mm mesh screen followed by cone and quarter splitting technique detailed in Schumacher et al. (1991) prior to wet packing into laboratory-based setups. Properties (i.e. uniformity coefficient (Cu), iron (Fe) and manganese (Mn)) of these media are presented in Table 4.2. Six samples (n=6) were analyzed for Fe, Mn and organic matter.

Table 4.2 Properties of the media used in laboratory-based columns

Media type	Size (mm)	Cu	Porosity (%)	Fe (mg/g soil)	Mn (mg/g soil)	Organic matter (µg TOC/g sand)
Silica sand	0.8-1.25	1.3	40	4.9±0.6 (n=6)	0.5±0.0 (n=6)	84.5±5.9 (n=6)
Dune sand	0.15-0.3	1.4	35.7	2.4±0.5 (n=6)	0.1±0.0 (n=6)	194.8±5.4 (n=6)

4.2.3 Experimental setup

Three uPVC columns were roughened from inside to minimize preferential and column interface flow and wet-packed in layers by allowing soil grains to settle in deionized water while slightly striking the surface of the column using a mallet rubber hammer to ensure homogeneous media packing in the column. Each column consisted of two parts, each 2.5 m high and an internal diameter of 57 mm. Both parts were connected in series and run in a down flow (gravity) mode by connecting the bottom of the first part to the top of the second part using a 5 mm diameter plastic tube (Tygon, Saint-Gobain Corporation). A ponding head space of 20 cm was provided on the top of each column and influent samples were taken from a port situated in this height to account for any PE quality change in connection tubes between the feed tank and column influent. Sampling ports were fitted at an interval of 25 cm in the first one meter along the column depth after which the interval between each successive ports was 50 cm. The ports were extended to the middle of the column using glass tubes of 5 mm diameter. Two columns were packed with silica sand and used to explore the impact of hydraulic loading rate, while the third column was packed with dune sand and used to assess the influence of media type on the removal of bulk organic matter, ammonium-nitrogen (NH_4-N) and nitrate-nitrogen (NO_3-N). Support layer of gravel with grain size ranging from 2 mm to 10 mm was used at the bottom of each column. A variable-speed peristaltic pump was used to continuously deliver the PE at constant HLRs of 0.625 m/d and 1.25 m/d resulting in empty bed contact times (EBCTs) of 8 and 4 days, respectively. These HLRs were frequently checked at influent and effluent points of each column using a measuring cylinder and a stopwatch. Biofilms formation around the media and subsequent bio-stability of the soil columns was monitored by analyzing dissolved organic carbon (DOC) for influent and effluent samples filtered through 25 mm diameter regenerated cellulose filter with nominal size of 0.45 µm (Whatman, Germany). Columns bio-stability was achieved after 60 days of operation when a difference of less than ±1% DOC removal was obtained between each (three) successive samples. Moreover, a period of 18-21 days was allowed after changing operating conditions to adapt the microorganisms to the new environmental conditions. Consequently, samples (n=3) were taken from various sampling ports fitted along the column profile and analyzed for DOC, NH_4-N and NO_3-N. Figure 4.1 presents schematic of the column setup used.

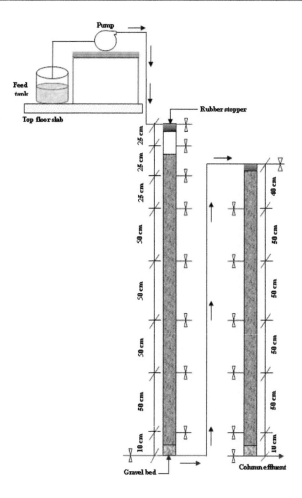

Figure 4.1 Schematic of soil column experimental setup

4.2.4   Analytical methods

Prior to soil column packing, concentrations of Fe and Mn of silica sand and dune sand were analyzed following the procedures outlined in Eaton et al. (2005). 5 g sample was submerged in a mixed solution containing 10 mL of nitric acid and 50 mL of Milli-Q water in a conical flask covered with a funnel and digested on a hot plate at $200\pm20°C$ until a volume of 10 ± 2 mL was reached. The funnel surface was rinsed into the solution and the sand and solution were transferred into a 50 mL flask and filled to the mark using Milli-Q water and left overnight. Three replicates of blank samples containing only Milli-Q water and acid were digested alongside with samples. Quantification of iron and manganese was then conducted using AAnalystTM 200 Atomic Absorption Spectrometer (PerkinElmer, USA). Organic matter content of the media was obtained by digesting 10 g of silica sand or dune soil with 50 mL Milli-Q water and 1 mL nitric acid for 30 minutes. The solution was then cooled down and soil organic matter was then measured as TOC in unfiltered samples. TOC of the used media, DOC concentrations and Ultraviolet absorbance at 254 nm wavelength (UVA_{254}) of all pre-filtered samples collected from WWTP and laboratory-based set-ups were determined in the laboratory within three days as detailed in section 3.2.4. Specific ultra violet light absorbance (SUVA) was calculated to explore the contribution of aromatic structures of DOC of the samples using their respective UVA_{254} measurements and DOC values. Fluorescence excitation-emission matrix (F-EEM) is used in water samples analysis (Baker, 2001). Since DOC contains chromophoric (light absorbing) and fluorophoric (light emitting) molecules, fluorescence excitation-emission matrices (F-EEM) are used in water samples analysis (Baker, 2001) to distinguish between different characteristic peaks depicting different DOC fractions. DOC concentrations of all samples were diluted with ultra-pure (Milli-Q) water to obtain 1 mg/L DOC concentration without pH adjustment. F-EEM spectra were then obtained through collection of a series of emission spectra at different excitation wavelengths using a FluoroMax-3 spectrofluorometer (HORIBA Jobin Yvon, Edison, NJ, USA). The area under the Raman scatter peak (at excitation wavelength of 350 nm) of Milli-Q water sample was used to calibrate fluorescence spectra, followed by removal of the Raman signal through subtraction of normalized Milli-Q EEM (Stedmon et al., 2003). In order to account for earlier DOC dilution, an order of magnitude at which the DOC was diluted (dilution factor) was used to carry out correction of fluorescence intensities in MATLAB (version 7.9, R2009b) used to illustrate organic matter fractions of humic-, fulvic- and protein-like as identified by Amy and Drewes (2007). These fractions were identified in MATLAB contour maps as peaks of an excitation-emission (Y-X) matrix. These peaks were referred to by a combination of letters and numbers. Peak 1 (P1) was assigned to (primary humic) humic-like, peak 2 (P2) was given to (secondary humic) fulvic-like while peak 3 (P3) stood for protein-like.

Chemical reagents used to measure NH_4-N and NO_3-N were of analytical grade and were purchased from Merck KGaA, Germany. NH_4-N and NO_3-N were determined according to the methods detailed under section 3.2.4. Measurement of DO was carried out using an HQ30d meter and LDO101 probe (Hach, Colorado, USA). pH of all samples was determined using 691 pH Meter (Metrohm, USA).

4.3   RESULTS AND DISCUSSION

4.3.1   Impact of hydraulic loading rate

4.3.1.1 Bulk organic matter

DOC removal of 47.5±1.2% (from 38.5±1.2 mg/L to 20.2±2.7 mg/L) and 45.1±1.2% (from 39.0±1.1 mg/L to 21.4±0.4 mg/L) was achieved in biologically stable soil columns operated at HLRs of 0.625 and 1.25 m/day, respectively. As shown in Figure 4.2, DOC removal at both HLRs remained nearly equal in the first 1.5 m and was presumably dominated by biodegradation. Nevertheless, the impact of long residence time of water at a HLR of 0.625 m/d increased DOC removal along the column causing it to diverge to 2.5 m. Between 3 and 5 m depth of the column, a slight decrease in DOC concentration was observed suggesting prevalence of adsorption in this part of the column. Biodegradation and adsorption mechanisms occur concurrently during SAT and part of the DOC adsorbed onto the media undergoes biodegradation (Idelovitch et al., 2003). This successive DOC removal occurs in most of the biological filters. Nevertheless, it is difficult to precisely quantify the amount of DOC removed by each mechanism.

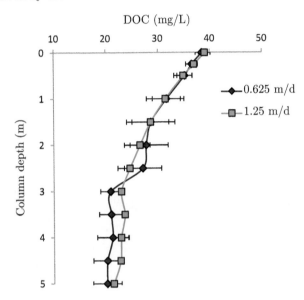

Figure 4.2 Change in DOC concentration along the depth of soil column at different HLRs (silica sand size: 0.8–1.25 mm)

SUVA steadily increased from 2.2±0.5 L/mg. m at the influent to 2.9±0.7 L/mg. m at the depth of 3 m, then decreased marginally to 2.8±0.7 L/mg. m at the effluent of the soil column operated at HLR of 0.625 m/d. Likewise, SUVA values exhibited an increase from 1.6±0.2 L/mg. m at the influent to 2.0±0.2 L/mg. m at column depth of 3 m followed by insignificant decrease to 1.9±0.2 L/mg. m at the effluent of the soil column operated at HLR of 1.25 m/d. Biodegradation of lower molecular weight DOC substances, release of high molecular weight soluble microbial products (SMPs) and leaching soil organic contribute to increase in SUVA (Westerhoff and Pinney, 2000). Conversely, reduction in SUVA values of effluent samples suggests preferential removal of higher molecular weight, hydrophobic and aromatic DOC constituents through sorption (Westerhoff and Pinney, 2000). Reduction in SUVA values between 3 to 5 m denotes that adsorption was the dominant removal mechanism for DOC.

Fluorescence gives relevant information about organic matter characteristics which are utilized to get more insight on organic matter structures (Baker et al., 2007).

Figure 4.3 shows F-EEM spectra of samples taken from influent and effluent points of the soil columns operated at 0.625 mg/L (a and b) and 1.25 m/d (c and d) were identified at different excitation-emission wavelengths. P1 ($\lambda_{ex/em}$ = 240-250/430 nm), P2 ($\lambda_{ex/em}$ = 330/415-425 nm), and P3 ($\lambda_{ex/em}$ = 270-280/310-330 nm) were observed.

(a)   (b)

(c)   (d)

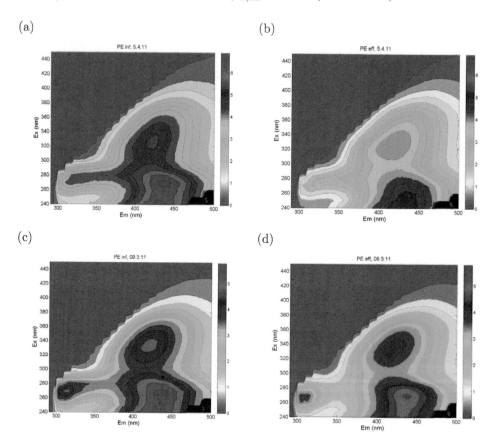

Figure 4.3 F-EEM spectra of influent and effluent at different HLR (a) influent 0.625 m/d (b) effluent 0.625 m/d (c) influent 1.25 m/d (d) effluent 1.25 m/d (silica sand size: 0.8–1.25 mm)

Change in fluorescence intensity was obtained by comparing fluorescence intensity of an effluent sample to that of the corresponding influent sample at the same range of excitation-emission wavelength. P1 was reduced by 22.8 and 18% in soil columns operated at HLRs of 0.625 and 1.25 m/d, respectively, while intensities of P2 showed reduction by 21.7% and 14.8% at similar HLRs. Furthermore, P3 was reduced by 47.7% at HLR of 0.625 m/d compared to 24.5% at HLR of 1.25 m/d. In general, humic substances resist biodegradation due to their hydrophobicity, but they could be removed through adsorption in the subsurface environment (Quanrud et al., 1996). Though the reduction of DOC fractions of P1 and P2 were comparable at both HLRs, substantial reductions in fluorescence intensities of P3 in the soil column operated at HLR of 0.625 m/d compared to those of 1.25 m/d suggested that biotransformation of dissolved organic matter (DOM) was higher in the columns operated at HLR of 0.625 m/d. However, the relatively high reduction of P1 and P2 at HLR of 0.625 m/d could be attributed to possibly low competition for adsorption sites due to biodegradation of organic matter and/or adsorption of these substances due to long residence time which allows these materials to migrate to adsorption sites.

4.3.1.2 Nitrogen

Influent NH_4-N concentrations entering columns operated at HLRs of 0.625 and 1.25 m/d were 31.9±2.1 and 28.6±7.1 mg N/L, respectively. The corresponding concentrations of NH_4-N exiting the columns were 8.3±1.2 mg N/L and 17.6±9.1 mg N/L. Average removals of NH_4-N were 74.5±18.0 and 39.2±4.3% at HLRs 0.625 and 1.25 m/d, respectively. As shown in Figure 4.4, NH_4-N concentration decreased considerably along the column depth at HLR of 0.625 m/d. However, NH_4-N concentration slightly increased in the first meter along the column operated at HLR of 1.25 m/d after which it decreased.

(a) (b)

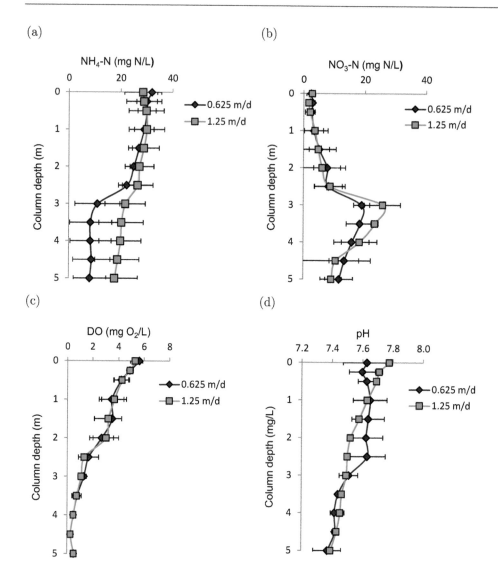

Figure 4.4 Change in (a) NH$_4$-N, (b) NO$_3$-N, (c) DO and (d) pH along the depth of soil columns at different HLRs (silica sand size: 0.8–1.25 mm)

Increase in NH$_4$-N concentration in the first meter along the 1.25 m/d column could be due to loading of NH$_4$-N at rates higher than its nitrification and desorption of the previously adsorbed NH$_4$-N. Ammonification of NH$_4$-N was excluded since the pH of the water was below 8.3. NO$_3$-N removal patterns showed an increase in the first 3 m along the column depth due to nitrification of NH$_4$-N. NO$_3$-N concentration increased from 2.4±1.4 mg N/L at the influent to 19.0±2.6 mg N/L at depth of 3 m and then decreased to 11.6±4.5 mg N/L at the effluent for the HLR of 0.625 m/d. On the other hand, experiments conducted at HLR of 1.25 m/d revealed increase in NO$_3$-N concentration from 2.7±0.8 mg N/L at the influent to 25.8±5.9 mg N/L at 3 m deep in the column, then decreased to 9.1±5.2 mg N/L in effluent samples. Results from HLR of 0.625 m/d column exhibited higher overall removal of NH_4-N compared to that at 1.25 m/d suggesting NH_4-N reduction through adsorption to the media at lower HLR due to long hydraulic residence time at HLR of 0.625 m/d which resulted in adsorption of NH_4-N to the soil alongside nitrification. Additionally, a substantial increase in NO_3-N was observed in the first 3 m of the soil column operated at HLR of 1.25 m/d as compared to NO_3-N concentrations at the same depth at the HLR of 0.625 m/d column. This could be ascribed to availability of more NH_4-N for nitrification due to relatively short hydraulic residence time and subsequently low adsorption or contribution from other nitrogen sources. However, it is not possible to differentiate between the NH_4-N removed through adsorption and that removed by nitrification (Idelovitch et al., 2003). Attenuation of DOC and NH_4-N is a biologically mediated process in which DO is utilized by microorganisms as an electron acceptor. DO (c) concentration followed a reduction pattern along the soil depth similar to that of DOC and NH_4-N. Microorganisms follow the route with the highest energy yield to achieve maximum cell synthesis (Essandoh et al., 2011). Low DO concentration and high oxygen demand of the primary effluent fed to the column promoted anoxic conditions at the depth of 3 m in both columns depicted by NO_3-N reduction through denitrification between 3 m and 5 m. Denitirification is an anoxic process in which microorganisms use NO_3-N as an electron acceptor. Denitrification in presence of DO known among microbiologists as co-metabolism of DO and NO_3-N (Gao et al., 2009) is a synchronous process which occurs when microorganisms use branches of their electron transport chain to direct electron flow simultaneously to denitrifying enzymes alongside DO (Chen et al., 2003). Reduction pattern of NO_3-N concentration deep in the column (4 m and 5 m) in presence of DO (0.2 to 0.5 mg/L) suggested that aerobic denitrification might have taken place in this part of the column.

4.3.2   Impact of soil type

4.3.2.1 Bulk organic matter

Soil type is considered as one of the factors that affects SAT efficiency and determines the extent of improvement in the reclaimed water quality. Figure 4.5 presents DOC attenuation along two soil columns packed with silica sand and dune sand and operated at HLR of 1.25 m/d. While the silica sand column showed DOC reduction of 45.1±1.2% (39.0±1.1 mg/L to 21.4±0.4 mg/L) dune sand column exhibited DOC removal of 57.3±7.6% (32.8±0.6 mg/L to 14.0±2.3 mg/L) out of which 37.1±4.2 and 47±3.4% took was removed in the first 2.5 m along the silica sand and dune sand columns, respectively.

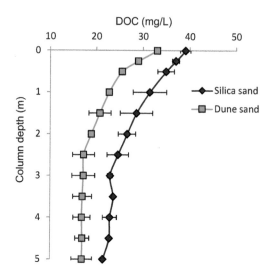

Figure 4.5 Change in DOC concentrations along the depth of soil columns packed with silica sand and dune sand (HLR = 1.25 m/d)

SUVA values increased in silica sand column from 1.6±0.2 L/mg. m in the influent to 2.0±0.2 L/mg. m then marginally decreased to 1.9±0.2 L/mg. m in the effluent. The same trend was observed in the dune sand column where SUVA increased from 2.3±0.2 to 4.8±1.0 L/mg. m and decrease to 4.7±0.8 L/mg. m. High removal of DOC and corresponding increase of SUVA along the depth of both columns is presumably due to prevalence of high biodegradation that preferentially removed aliphatic (readily biodegradable organic) substances (Cha et al., 2004). Despite the apparent high DOC removal in dune sand column, comparable DOC amounts of 17.6±0.9 and 19.1±3.0 mg/L were removed in silica sand and dune sand columns, respectively. DOC removal in both columns appeared to be load dependent since dune sand column received less DOC load (32.8±0.6 mg/L) while the removal capacity of the silica sand column was affected by a relatively high DOC load (39.0±1.1 mg/L).

F-EEM spectra of samples collected from silica sand and dune sand column experiments exhibited contour maps showing the three main peaks detected. Excitation-emission wavelengths at which these peaks were identified are: P1 ($\lambda_{ex/em}$ = 240-250/430-432 nm), P2 ($\lambda_{ex/em}$ = 320-340/420-422 nm) and P3 ($\lambda_{ex/em}$ = 270-280/314-350 nm). A comparison of influent and effluent samples from silica sand and dune sand column experiments operated at HLR of 1.25 m/d revealed reductions in fluorescence intensity of P1 by 18 and 21.4% in silica sand and dune sand, respectively, while P2 showed reduction by 14.8 and 25.7% in both columns. Furthermore, fluorescence intensities of P3 were reduced by 24.5 and 39.3% in silica sand and dune sand columns, respectively. Figure 4.6 shows F-EEM spectra for silica and dune sand columns.

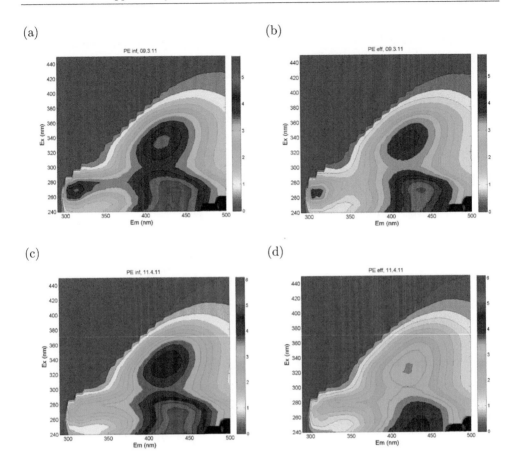

Figure 4.6 F-EEM spectra of influent and effluent of soil columns packed with different media (a) influent silica sand (b) effluent silica sand (c) influent dune sand and (d) effluent dune sand (aerobic conditions HLR = 1.25 m/d, EBCT = 4 days).

Despite of the high soil organic matter (SOM) content of dune sand, P1 and P2 and P3 were fairly significantly reduced in dune sand column compared to silica sand column. While humic fractions of organic carbon (P1 and P2) resist biodegradation due to their hydrophobicity, their removal in the subsurface environment is achieved through adsorption (Quanrud et al., 1996). However, protein-like fractions are biologically removed. High removal of P1, P2 and P3 in dune sand column was influenced by low DOC load.

4.3.2.2 Nitrogen

Figure 4.7 shows change in (a) NH_4-N, (b) NO_3-N, (c) DO, and (d) pH along silica sand and dune sand columns. Total NH_4-N removal of 39.1±4.3% (28.6±7.1 mg/L to 17.6±9.1 mg/L) in soil column packed with silica sand while a removal of 49.2±5.2% (25.1±0.6 mg/L to 12.7±1.0 mg/L) was attained in dune sand column. Profiles of DO

80

along the column depth showed that DO was substantially removed in the first 2 m in the column packed with dune sand. However, average DO concentration in effluent samples in both columns was 0.5 mg/L.

(a)

(b)

(c)

(d)

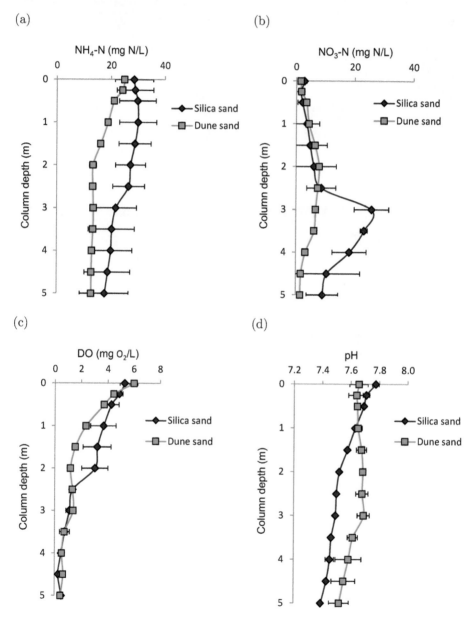

Figure 4.7 Change in (a) NH$_4$-N, (b) NO$_3$-N, (c) DO and (d) pH along the depth of silica sand and dune sand columns (HLR = 1.25 m/d)

Reduction in NH_4-N concentration had triggered increase in NO_3-N concentration along the column depth through nitrification and bio-adsorption processes from 2.7 ± 0.8 mg/L (h = 0 m) to 25.8 ± 5.9 mg/L (h = 3 m) followed by decrease to 9.1 ± 5.2 mg/L (h = 5 m). On the other hand, NO_3-N increased from 1.5 ± 0.2 mg/L (h = 0 m) to 7.8 ± 0.6 mg/L (h = 2 m) and significantly decreased to 1.5 ± 0.5 mg/L at the effluent of soil column packed with dune sand. Even though NH_4-N was removed in dune sand column at rate greater (10.1%) than that in the silica sand column, it did not necessarily translate into high NO_3-N generation in the first 3 m deep in the column. On the contrary, the silica sand column showed corresponding high increase in NO_3-N concentration compared to dune sand column. Since anoxic removal of ammonium in presence of organic carbon has been reported (Sabumon, 2007), high removal of NH_4-N in the dune sand material column might be attributed to anaerobic ammonium oxidation (ANAMMOX) and denitrification leading to less NO_3-N concentration detected at the effluent of the column. Low porosity of dune sand and its small texture might have enhanced the growth of diverse microbial community in the biofilms environment around media grains and caused localized anoxic zones in the biofilms in which NH_4-N was converted to nitrogen (N_2).

4.4   CONCLUSIONS

DOC removal was not dependent on the HLR at which the soil column experiments were operated in the first 3 m of the column. However, soil column operated at HLR of 0.625 m/d exhibited DOC removal higher than what was achieved in soil column operated at HLR of 1.25 m/d between 3 to 5 m through adsorption due to long residence time at HLR of 0.625 m/d.

Fluorescence intensities (P1, P2 and P3) were considerably reduced at HLR of 0.625 m/d column by 22.8%, 21.7% and 47.7% compared to 18.0%, 14.8% and 24.5% in 1.25 m/d column suggesting that long hydraulic retention time influenced the removal of these intensities in silica sand.

Column experiments carried out at 1.25 m/d using silica sand and dune sand revealed minor difference in the DOC   removal   implying   that DOC removal was not dependent on the media used, but on rather on the DOC load applied to the column. NH_4-N was reduced by $74.5\pm18.0\%$ and $39.2\pm4.3\%$ through nitrification and adsorption mechanisms at HLR of 0.625 m/d and 1.25 m/d, respectively. The first 3 m of the soil column was dominated by nitrification whereas the reduction in the last 2 m was dominated by adsorption leading to higher NH_4-N reduction at HLR of 0.625 m/d.

Removal of NO_3-N between 3 and 5 m along the column in the presence of low dissolved oxygen concentration suggested that denitrification might have taken place in that part of the column due to presence of anoxic pockets in the medium or localized anoxic spots in the biofilms following depletion of DO.

In general, reduction of DOC and NH_4-N was not much dependent on HLR in the first 3 m of soil column as it was dominated by biological oxidation mechanisms (biodegradation and nitrification). However, DOC, NH_4-N and NO_3-N reduction from PE was achieved through adsorption and denitrification in the last 2 m of the columns (3 to 5 m).

Results of this study show that while DOC removal from PE in SAT system is not dependent on soil type or HLR, DOC removal in two different soils is likely to be higher in the one with finer particles which provides more surface area for biofilms to develop. Furthermore, the findings suggest that NH_4-N reduction in SAT is likely to be relatively higher under low HLR and soil with fine particles due to longer contact time and presence of adsorption binding sites. The practical implication of using such HLR and soil type is the need for much frequent drying and scraping of soil surface to remove any clogging layer.

4.5   REFERENCES

Amy, G. L. and Drewes, J. (2007). Soil aquifer treatment (SAT) as a natural and sustainable wastewater reclamation/reuse technology: Fate of wastewater effluent organic matter (EfOM) and trace organic compounds. *Environmental Monitoring and Assessment,* **129**(1), 19-26.

Baker, A. (2001). Fluorescence excitation-emission matrix characterization of some sewage-impacted rivers. *Environmental Science and Technology,* **35**(5), 948-953.

Baker, A., Elliott, S. and Lead, J. (2007). Effects of filtration and pH perturbation on freshwater organic matter fluorescence. *Chemosphere,* **67**(10), 2035-2043.

Bouwer, H. (1991). Role of groundwater recharge in treatment and storage of wastewater for reuse. *Water Science and Technology,* **24**(9), 295-302.

Cha, W., Choi, H., Kim, J. and Kim, I. (2004). Evaluation of wastewater effluents for soil aquifer treatment in South Korea. *Water Science and Technology,* **50**(2), 315-322.

Chen, F., Xia, Q. and Ju, L. K. (2003). Aerobic denitrification of Pseudomonas aeruginosa monitored by online NAD (P) H fluorescence. *Applied and Environmental Microbiology,* **69**(11), 6715-6722.

Crites, R. W., Reed, S. C. and Middlebrooks, E. J. (2006). *Natural Wastewater Treatment Systems.* CRC Press, Boca Raton, Florida, USA, pp 413-426.

Dillon, P., Kumar, A., Kookana, R., Leijs, R., Reed, D., Parsons, S. and Ingleton, G. (2009a). Managed Aquifer Recharge - Risks to Groundwater Dependent Ecosystems: A Review. Land and Water, Canberra, Australia.

Dillon, P., Pavelic, P., Page, D., Beringen, H. and Ward, J. (2009b). *Managed Aquifer Recharge: An Introduction.* National Water Commission, Australia.

Drewes, J., Quanrud, D., Amy, G. and Westerhoff, P. (2006). Character of organic matter in soil-aquifer treatment systems. *Journal of Environmental Engineering,* **132**, 1447-1458.

Eaton, A. D., Clesceri, L. S., Rice, E. W. and Greenberg, A. E. (2005). *Standard Methods for the Examination of Water and Wastewater. 21th ed.* American

Public Health Association, American Water Works Association, and Water Environment Federation, Washington, DC, USA.

Emerick, R., Test, R., Tchobanoglous, G. and Darby, J. (1997). Shallow intermittent sand filtration: Microorganisms removal. *Small Flows Journal*, **3**(1), 12-22.

Essandoh, H. M. K., Tizaoui, C., Mohamed, M. H. A., Amy, G. and Brdjanovic, D. (2011). Soil aquifer treatment of artificial wastewater under saturated conditions. *Water Research*, **45**(14), 4211-4226.

Fox, P., Houston, S., Westerhoff, P., Drewes, J., Nellor, M., Yanko, B., Baird, R., Rincon, M., Arnold, R. and Lansey, K. (2001). An Investigation of Soil Aquifer Treatment for Sustainable Water Reuse. *Research Project Summary of the National Center for Sustainable Water Supply (NCSWS)*, Tempe, Arizona, USA.

Gao, H., Schreiber, F., Collins, G., Jensen, M. M., Kostka, J. E., Lavik, G., de Beer, D., Zhou, H. and Kuypers, M. M. M. (2009). Aerobic denitrification in permeable Wadden Sea sediments. *The ISME Journal*, **4**(3), 417-426.

Ho, G., Gibbs, R., Mathew, K. and Parker, W. (1992). Groundwater recharge of sewage effluent through amended sand. *Water Research*, **26**(3), 285-293.

Idelovitch, E., Icekson-Tal, N., Avraham, O. and Michail, M. (2003). The long-term performance of Soil Aquifer Treatment(SAT) for effluent reuse. *Water Science and Technology: Water Supply*, **3**(4), 239-246.

Nema, P., Ojha, C., Kumar, A. and Khanna, P. (2001). Techno-economic evaluation of soil-aquifer treatment using primary effluent at Ahmedabad, India. *Water Research*, **35**(9), 2179-2190.

Pescod, M. (1992). Wastewater Treatment and Use in Agriculture. Food and Agriculture Organization Irrigation and Drainage Paper 47. Rome

Quanrud, D., Arnold, R., Wilson, L. and Conklin, M. (1996). Effect of soil type on water quality improvement during soil aquifer treatment. *Water Science and Technology*, **33**(10), 419-432.

Rauch-Williams, T., Hoppe-Jones, C. and Drewes, J. E. (2010). The role of organic matter in the removal of emerging trace organic chemicals during managed aquifer recharge. *Water Research*, **44**(2), 449-460.

Rauch, T. and Drewes, J. (2004). Assessing the removal potential of soil-aquifer treatment systems for bulk organic matter. *Water Science and Technology*, **50**(2), 245-253.

Sabumon, P. C. (2007). Anaerobic ammonia removal in presence of organic matter: A novel route. *Journal of Hazardous Materials*, **149**(1), 49-59.

Schumacher, B., Shines, K., Burton, J. and Papp, M. (1991). A Comparison of Soil Sample Homogenization Techniques. *Hazardous Waste Measurements.*" MS Simmons, ed. Chelsea, Michigan: Lewis Publishers, pp. 53-68.

Sharma, S. K., Hussen, M. and Amy, G. L. (2011). Soil aquifer treatment using advanced primary effluent. *Water Science and Technology*, **64**(3), 640-646.

Stedmon, C., Markager, S. and Bro, R. (2003). Tracing dissolved organic matter in aquatic environments using a new approach to fluorescence spectroscopy. *Marine Chemistry*, **82**(3-4), 239-254.

Westerhoff, P. and Pinney, M. (2000). Dissolved organic carbon transformations during laboratory-scale groundwater recharge using lagoon-treated wastewater. *Waste Management,* **20**(1), 75-83.

Wilson, L., Amy, G., Gerba, C., Gordon, H., Johnson, B. and Miller, J. (1995). Water quality changes during soil aquifer treatment of tertiary effluent. *Water Environment Research,* **67**(3), 371-376.

Yamaguchi, T., Moldrup, P., Rolston, D. E., Ito, S. and Teranishi, S. (1996). Nitrification in porous media during rapid unsaturated water flow. *Water Research,* **30**(3), 531-540.

CHAPTER 5

INFLUENCE OF WETTING AND DRYING CYCLES ON REMOVAL OF SUSPENDED SOLIDS, BULK ORGANIC MATTER, NUTRIENTS AND PATHOGENS INDICATORS FROM PRIMARY EFFLUENT IN MANAGED AQUIFER RECHARGE[3]

SUMMARY

The impact of intermittent application of primary effluent on the removal of suspended solids, bulk organic matter, nitrogen and pathogens indicators during soil aquifer treatment was investigated using 4.2 m high laboratory-scale soil columns fitted with sampling ports along the depth. Continuous and intermittent modes of application were adopted using peristaltic pumps to deliver the primary effluent to the columns operated at hydraulic loading rates of 0.625 and 1.25 m/d with varying wetting and drying periods. Experimental results exhibited insignificant change in suspended solids and dissolved organic carbon removals under continuous and intermittent mode of primary effluent application. While the overall removal of suspended solids ranged from 86 to 95%, the overall removal of dissolved organic carbon ranged from 50 to 60% irrespective of the length of wetting and drying period or the hydraulic loading rate. Nevertheless, reduction of ammonium-nitrogen varied significantly with the length of drying period and the highest reductions of 88.4±0.8 and 98.0±0.1% were achieved at 3.2 days wetting/6.4 days drying and 6.4 days wetting/6.4 days drying, respectively. Likewise, the removal of *E. coli* and *total coliforms* increased significantly with the increase in the drying period resulting in more than 4 $\log_{10}$ units under similar operating conditions. These results suggest that while removals of suspended solids and dissolved organic carbon were independent of mode of effluent application, removals of nitrogen and pathogens indicators were dependent on the length of the drying cycle.

---

[3] Based on Abel et al. (2014). *Ecological Engineering*, **64**, 100-107.

5.1 INTRODUCTION

Interaction of hydraulic and purification processes during soil aquifer treatment (SAT) leads to formation of a clogging layer (van Cuyk et al., 2001) which in turn causes reduction of infiltration rates to as little as 10% of their original rates (Greskowiak et al., 2005). Soil clogging is controlled by periodic drying of infiltration facility to allow the clogging layer to decompose, shrink, crack and curl up (Bouwer, 2002) followed by application of wastewater effluent. This cyclic operation of wetting and drying helps in restoration of infiltration rates and allows oxygen to diffuse through the clogging layer to the underlying soil. Ammonium is adsorbed onto the soil media during wetting cycles whereas drying cycles increase aeration of the soil beneath the recharge facility which is utilized by nitrifying bacteria to oxidize the adsorbed ammonium (NRC, 2012). Furthermore, removal of easily biodegradable organic carbon in the infiltration zone causes depletion of oxygen and consequently changes the redox conditions to anoxic (NRC, 2012). Several full- and laboratory-scale SAT systems have used wetting/drying cycles that vary in the length of wetting or drying period, ranging from wetting periods longer/shorter than drying periods to wetting periods equal to drying periods (Kopchynski et al., 1996; Pescod, 1992). Generally, a short wetting period of less than 7 days is sufficient to prevent ammonium ion from breaking through sub-surface soils, whilst drying period should be as long enough (greater/equal to 4 days for coarser soils) to enable oxygen to penetrate deep in the soil where it is utilized in biological oxidation of ammonium ions (Fox et al., 2001). However, operating conditions must be based on local site characteristics and weather patterns since they are influenced by environmental factors including temperature, precipitation and solar incidence (Fox et al., 2001). SAT systems have been employed to treat wastewater effluents for a range of reuse purposes. However, most of these systems use secondary or tertiary effluents and limited work has been done to demonstrate the applicability and suitability of SAT in further polishing of primary effluents in developing countries where sophisticated wastewater treatment systems are not widely used due to lack of financial resources (Sharma et al., 2011).

The objective of this study was to study the effect of wetting and drying cycles on the removal of bulk organic matter, nitrogen, suspended solids and pathogens indicators from primary effluent (PE) in a laboratory-scale SAT system.

5.2 MATERIALS AND METHODS

5.2.1 Source water characteristics

Source and main characteristics of PE are similar to that used for probing the effect of PE pre-treatment on removal of bulk organic matter, nitrogen and pathogen indicators detailed under section 3.2.1.

5.2.2 Experimental setup

Two typical uPVC columns were wet-packed with silica sand (grain size 0.8 - 1.25 mm). A ponding headspace of 20 cm was provided on the top of each column and influent samples were taken from a port situated in this length to account for any PE quality change in the connection tubes between the feed tank and the column headspace. Dimensions of the columns used and spacing between various sampling ports along the column depth are presented in section 3.2.3.

A variable-speed peristaltic pump was used to continuously deliver the PE to the top of the columns at hydraulic loading rates (HLRs) of 0.625 m/d and 1.25 m/d at room temperature. Hydraulic residence time (HRT) of 6.4 and 3.2 days were obtained under continuous PE loading (wetting) at HLRs of 0.625 and 1.25 m/d, respectively. Nevertheless, frequent drying of the columns was subsequently adopted to mimic typical SAT field operating conditions. While wetting/drying cycles used at HLR of 0.625 m/d were 6.4 days wetting/ 1 day drying, 6.4 days wetting/ 3.2 day drying (1:0.5) and 6.4 days wetting/ 6.4 days drying (1:1), wetting and drying periods applied at HLR of 1.25 m/d were 3.2 days wetting/ 1 day drying, 3.2 days wetting/3.2 days drying (1:1) and 3.2 days wetting/6.4 days drying (0.5:1). Out of the drying periods, 0.32 and 0.16 day constituted the time spent to drain PE from the ponding headspace (20 cm) at HLRs of 0.625 and 1.25 m/d, respectively. Nevertheless, drying depths achieved in the column operated at HLR of 0.625 m/d were 0.4, 1.8 and 3.8 m for drying periods of 1, 3.2 and 6.4 days, respectively. Furthermore, 1.1, 3.8 and >4 m drying depths were attained under identical drying periods at HLR of 1.25 m/d.

Biofilms formation on the media and subsequent bio-stability of the soil columns was monitored for 80 days (during continuous wetting) by analyzing dissolved organic carbon (DOC) of influent and effluent samples filtered through 25 mm diameter regenerated cellulose filter with nominal size of 0.45 μm (Whatman, Germany) after which DOC removal was calculated. The columns were assumed to be ripened (bio-stablilized) when a difference of ±1 % DOC removal was obtained between each successive pair of samples. Furthermore, ripening process was repeatedly carried out for 3-5 weeks when wetting and drying conditions of the columns were changed to ensure biomass acclimation to the new environmental conditions.

5.2.3 Analytical methods

Analytical methods used to analyze suspended solids (SS), dissolved organic carbon (DOC), nitrogen and pathogens indicators in this chapter are similar to that used for analysis of impact of pre-treatment of PE on removal of SS, DOC, fluorescence excitation-emission matrices (F-EEM), nitrogen and pathogens indicators are similar to the methods elaborated under sections 3.2.4 and 4.2.4.

5.3 RESULTS AND DISCUSSION

5.3.1 Wetting and drying at HLR of 0.625 m/d

5.3.1.1 Suspended solids

Samples collected from the upmost part of soil columns operated at HLR of 0.625 m/d at different wetting/drying cycles exhibited relatively high SS removal (>70%) from PE. Removals of SS achieved at the depth of 50 cm along the column depth were 73.6±5.6, 72.7±3.2, 72.3±8.7 and 89.7±0.0% under continuous wetting, 6.4 days wetting/1 day drying, 6.4 days wetting/3.2 days drying and 6.4 days wetting/6.4 days drying, respectively. However, a slight improvement in SS removal was achieved further down in the column where total removals of 88.5±3.5% (effluent SS = 15.5±2.1 mg/L), 86.8±2.0% (effluent SS = 16.1±0.4 mg mg/L), 90.6±1.9% (effluent SS = 12.1±0.4 mg/L) and 95.2±0.0% (effluent SS = 6.0±1.9 mg/L) were observed at columns outlets under the same operating conditions. The overall SS removal under different mode of PE application showed no significant difference ($P = 0.06$) and was not dependent on the drying period since most of the SS was sieved out in the upper 50 cm of the column.

5.3.1.2 Bulk organic matter

Figure 5.1 presents, for continuous wetting and different wetting/drying cycles, average DOC concentrations (n=3) along the column depth. DOC was primarily removed within the top 50 cm of the column by 36.9±2.4, 47.3±9.4, 39.1±3.7 and 58.4±2.9% for continuous wetting, 6.4 days wetting/1 day drying, 6.4 days wetting/3.2 days drying and 6.4 days wetting/6.4 days drying, respectively. However, there was no significant variation ($P = 0.75$) in the overall DOC removal under various operating conditions. While DOC removal of 57.4±5.4% (effluent DOC = 14.1±2.2 mg/L) was achieved under continuous application of PE, one 6.4 days wetting coupled with one day drying resulted in DOC removal of 61.5±2.9% (effluent DOC = 14.7±1.9 mg/L). Furthermore, DOC removal of 57.4±1.8% (effluent DOC = 15.2±0.2 mg/L) was attained when 6.4 day wetting/3.2 days drying was adopted. On the other hand, equal wetting/drying periods of the column resulted in total DOC removal of 60.0±2.5% (effluent DOC = 12.9±0.0 mg/L). Aerobic degradation of DOC resulted in increase of SUVA values by 32.2±2.2% in the top 50 cm from average values of 1.5 L/mg. m at the influent to 2.9 L/mg. m at the depth of 50 cm indicating dominance of biological (biodegradation) mechanisms and removal of aliphatic DOC fractions (Quanrud et al., 2003). SUVA values then steadily decreased by 94.3±1.3% at the depths below 50 cm along the column to 2.0 L/mg. m at the outlet of the columns due to removal of aromatic DOC components with high molecular weight in that part of the column implying dominance of physical (adsorption) removal mechanisms (Gruenheid et al., 2005). Removal of both aliphatic and aromatic DOC substances occurs concurrently in the top part of the column. However, adsorption of the aromatic fractions in this part of the column was presumably reduced by presence of relatively high concentration of phosphorus

(7.8±2.8 mg/L) in PE which competed with aromatic substances for adsorption sites. The observed successive increase ($P = 0.225$) and decrease ($P = 0.742$) in SUVA values along soil columns were found to be independent of the length of the drying period. DOC removal at SAT infiltration zone causes depletion of dissolved oxygen (DO) and consequently promotes anoxic conditions (NRC, 2012). DOC removal achieved during application of 6.4 days wetting/6.4 days drying cycle was notably higher at the upper 50 cm of the column compared to continuous wetting and 6.4 days wetting/3.2 days drying, presumably due to aeration of the top part of the media leading to rapid DOC biodegradation. However, comparative overall DOC removal was achieved during continuous wetting and 6.4 days wetting/3.2 days drying at 50 cm due to competition between heterotrophic bacteria and ammonium oxidizing bacteria for DO leading to limited DOC reduction. A pronounced DOC removal was achieved further down in these columns, presumably through adsorption of DOC as these operating conditions preceded 6.4 days wetting/6.4 days drying cycle.

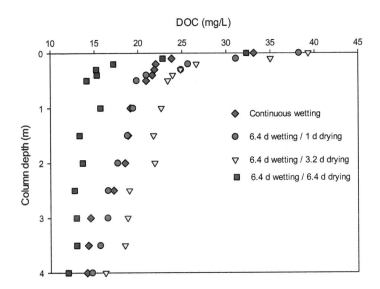

Figure 5.1 Average DOC concentration as a function of soil column depth (influent: primary effluent, media size: 0.8–1.25 mm, HLR = 0.625 m/d)

Fluorescence intensities were identified in three regions in 3D F-EEM spectra. These intensities were differentiated based on the range of excitation and emission wavelengths at which they occurred. Humic-like peak (P1) was observed in wavelength range of ($\lambda_{Ex/Em}$ = 240-250/425-445 nm), fulvic-like peak (P2) covered wavelength range of ($\lambda_{Ex/Em}$ = 325-330/424-428 nm) and protein-like peak (P3) appeared at the wavelength range of ($\lambda_{Ex/Em}$ = 270-280/340-350 nm). P1 was reduced by 22.2% during continuous application of PE for 6.4 days. This reduction increased to 24.8, 29.8 and 35.7% when wetting and drying cycles were introduced at 6.4 days wetting/1 day drying, 6.4 days wetting/3.2 days drying and 6.4 days wetting/6.4 days drying, respectively. Likewise, P2 reduction increased progressively with increase of the drying period and exhibited 19.4, 22.6, 26.1 and 29.3%. Nevertheless, comparison of influent and effluent P3 under different operating conditions revealed comparative biological reduction of P3 under continuous wetting, 6.4 days wetting/1 day drying and 6.4 days wet/3.2 days dry by 52.1, 50.1 and 55.1%, respectively. Furthermore, application of 6.4 days wetting/6.4 days drying showed P3 reduction of 61.7%. These results exhibited P3 reduction higher than P1 and P2 which is attributable to presence of high biodegradable DOC in the former (Maeng et al., 2012). Biodegradation of protein-like was better achieved at high drying period presumably due to prevalent aerobic conditions in the column operated at 6.4 days wetting/6.4 days drying. Table 5.1 presents F-EEM peaks reduction.

Table 5.1 Change in fluorescence peaks intensity during application of PE at different wetting/drying cycles (media size: 0.8–1.25 mm, HLR = 0.625 m/d, temperature: 20-22°C)

Operating condition	Intensity change in fluorescence peaks (%)		
	P1 ($\lambda_{Ex/Em}$ = 240-250/425-445 nm)	P2($\lambda_{Ex/Em}$ = 325-330/424-430 nm)	P3 ($\lambda_{Ex/Em}$ = 270-280/340-350 nm)
Continuous wetting	22.2	19.4	52.1
6.4 d wetting/ 1 d drying	24.8	22.6	50.1
6.4 d wetting/ 3.2 d drying	29.8	26.1	55.1
6.4 d wetting/ 6.4 d drying	35.7	29.3	61.7

5.3.1.3 Nitrogen

NH_4-N concentrations were monitored along the column profile. NH_4-N decreased notably in the first 50 cm with the maximum reduction (94.2±0.4%) achieved in the column with maximum drying period (6.4 days). Further significant (P = 0.002) overall NH_4-N reduction corresponded with increase in drying period. Total NH_4-N reduction of 36.8±7.6% (effluent 31.1±1.7 mg N/L), 45.2±0.1% (effluent 29.6 mg N/L), 59.3±4.3% (effluent 19.8±3.5 mg N/L) and 98.0±0.1% (effluent 0.8±0.0 mg N/L) was attained at effluent points of various soil columns operated at continuous wetting, 6.4 days wetting/1 day drying, 6.4 days wetting/3.2 days drying and 6.4 days wetting/6.4 days drying, respectively. NH_4-N reduction along the column profile corresponded with increase in NO_3-N concentration in the first 50 cm during which the latter increased from as low as 0.6±0.2 at the influent to 3.3±0.3 and 3.0±0.3 mg N/L at 50 cm under continuous wetting and 6.4 days wetting/3.2 days drying operating conditions, respectively. However, at equal wetting and drying periods of 6.4 days wetting/6.4 days drying, NO_3-N concentration increased remarkably from 1.0±0.1 to 20.7±0.4 mg N/L at 50 cm then decreased steadily to 12.8±0.3 mg N/L at the effluent. Biological reduction of NH_4-N was limited by availability of DO which decreased markedly in the first 50 cm from as much as 7.8±0.1 to as low as 0.5±0.0 mg O_2/L with the lowest DO concentration measured at the effluent of the column with the longest wetting/drying cycle. This concurrent NH_4-N reduction and NO_3-N increment in the first 50 cm along the flow path demonstrates that nitrification was the dominant NH_4-N reduction mechanism as evidenced by decrease in DO, while reduction in NO_3-N in the subsequent 3.5 m of the column suggests dominance of denitrification process sparked by low DO concentration. NH_4-N was reduced through cation exchange and sorbed onto soil grains during wetting cycles. However, the sorbed NH_4-N was further reduced through nitrification during drying cycle (NRC, 2012). Consistent NH_4-N reduction ($P = 0.016$) was achieved with drying conditions revealing further penetration of DO deeper in the column during drying periods which was presumably utilized by microorganisms to mediate biological reduction of DOC and NH_4-N during successive wetting cycles. Figure 5.2 shows NH_4-N and NO_3-N concentration profiles along the column depth at various wetting/drying cycles.

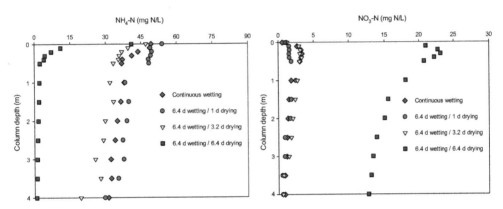

Figure 5.2 Average NH_4-N and NO_3-N concentrations profiles along the soil column depth operated at alternate wetting/drying cycles (influent: primary effluent, media size: 0.8–1.25 mm, HLR = 0.625 m/d)

5.3.1.4 E. coli and total coliforms removal

Removal of *E. coli* and *total coliforms* from PE was examined for a range of soil column wetting/drying conditions. Mean concentrations of indicators pathogens in influent PE fed to the setups was $6.2\times10^6\pm1.2\times10^6$ CFU/100 mL for *E. coli* and $22\times10^6\pm3.4\times10^6$ CFU/100 mL for *total coliforms*. Significant removal of *E. coli* and total coliforms in laboratory-scale and field MAR systems is well documented by many researchers (Abel et al., 2012; Hijnen et al., 2005; Icekson-Tal et al., 2003; Levantesi et al., 2010). Attenuation of pathogens is achieved through inactivation, straining and attachment to aquifer materials (McDowell-Boyer et al., 1986). *E. coli* reduction accounted to 2.8 ± 0.3, 3.1 ± 0.4, 3.4 ± 0.5 and 4.3 ± 0.5 log_{10} units under continuous wetting, 6.4 days wetting/1 days drying, 6.4 days wetting/3.2 days drying, respectively. Additionally, reduction of *total coliforms* under similar conditions was 2.6 ± 0.3, 2.8 ± 0.1, 3.5 ± 0.0 and 4.0 ± 0.1 log_{10} units. Similar removals for *E. coli* (3.1 log_{10}) and *total coliforms* (3.1 log_{10}) were achieved by Abel et al. (2012) during continuous infiltration of PE in a 30 cm column operated at HLR of 0.625 m/d.

Significant removal of *E. coli* ($P = 0.002$) and *total coliforms* ($P = 0.032$) was attained at longer drying cycles due to creation of more adsorption sites. Most of *E. coli* and *total coliforms* removal occurred in the upmost part of the column predominantly through adsorption and straining and longer drying periods have improved the removal of *E. coli* and *total coliforms* by allowing these pathogens indicators to penetrate deep down the column creating more adsorption sites where these species were adsorbed. Figure 5.3 shows removal of *E. coli* and *total coliforms* during infiltration of PE through soil column at different wetting/drying operating conditions.

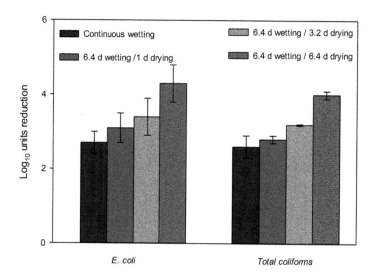

Figure 5.3 Pathogens indicators reduction during continuous and intermittent infiltration of primary effluent in soil columns (media size: 0.8–1.25 mm, HLR = 0.625 m/d)

5.3.2 Wetting and drying at HLR of 1.25 m/d

5.3.2.1 Suspended solids

There was a remarkable filtration of SS from PE in the first 50 cm of the column during which 75.3±6.4, 82.6±0.5, 73.4±2.8 and 77.9±1.1% was removed under continuous wetting, 3.2 days wetting/1 day drying, 3.2 days wetting/3.2 days drying and 3.2 days wetting/6.4 days drying, respectively. Furthermore, an improved removal of SS was observed in the effluent of soil columns under similar operating conditions where 90.9±2.7 (effluent SS = 9.0±3.0 mg/L), 91.2±1.4% (effluent SS = 9.7±0.7 mg/L), 88.3±0.3% (effluent SS = 15.5±2.1 mg/L) and 92.7±0.7% (effluent SS = 9.0±1.0 mg/L) SS was achieved. Comparable SS removal at different operating conditions suggests inconsistent ($P = 0.126$) correlation between SS removal and the length of the drying cycle. This shows that removal of SS occurred in the top part of the column irrespective to HLR or wetting/drying cycle.

5.3.2.2 Bulk organic matter

DOC removal in the top 50 cm of soil column varied considerably. Continuous PE application exhibited DOC removal of 35.2±4.4%, whereas 1 day drying of the column revealed DOC removal of 48.8±6.2%. Furthermore, DOC removal of 40.2±3.1% was achieved when the column was wetted and dried for equal periods (3.2 days) and 45.3±1.3% DOC removal was attained under 3.2 days wetting/6.4 days drying. However, DOC exiting the columns under similar conditions decreased by 49.4±0.5% (effluent DOC = 15.4±1.7 mg/L), 57.7±5.7% (effluent DOC = 13.8 mg/L), 58.1±2.2% (effluent DOC = 14.5±0.4 mg/L) and 57.1±0.5% (effluent DOC = 13.1±1.0 mg/L) under continuous, 3.2 days wetting/1 day drying, 3.2 days wetting/3.2 days drying and 3.2 days wetting/6.4 days drying, respectively. Corresponding SUVA values increased by 30.2±4.3 % from an influent value of 1.5±0.3 to 2.9±0.1 L/mg. m at 50 cm, and then decreased by 94.0±1.3% to 2.0±0.0 L/mg. m at the effluent. This increase in SUVA values is attributed to preferential removal of non-aromatic (aliphatic) near the surface of the soil. Conversely, reduction in SUVA values of effluent samples suggests preferential removal of higher molecular weight, hydrophobic and aromatic DOC constituents through sorption (Westerhoff and Pinney, 2000). DOC removal increased with introduction of wetting/drying cycles due to aeration of the top 50 cm of the media. Nevertheless, this increase in DOC removal was insignificant ($P = 0.948$). Likewise, increase of SUVA values ($P = 0.316$) in the upmost 50 cm of the column and subsequent decrease ($P = 0.949$) in the remaining 3.5 m of the column were insignificant when the drying period was increased. DOC profile along the depth of the column is plotted in Figure 5.4.

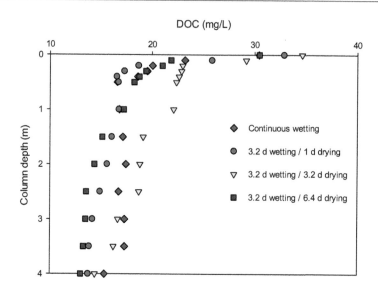

Figure 5.4 Average DOC concentrations as a function of soil column depth (influent: primary effluent, media size: 0.8–1.25 mm, HLR = 1.25 m/d)

F-EEM spectra of influent and effluent samples of the soil columns operated at HLR of 1.25 m/d under continuous and wet/dry cycles were analyzed at excitation-emission wavelengths similar to those observed at HLR of 0.625 m/d. P1 reduction through sorption increased with increase of the drying period from 12.99% under continuous wetting, 14.8% under 3.2 days wetting/1 day drying whereas 20.3 and 31% reductions were achieved during 3.2 days wetting/3.2 days drying and 3.2 days wetting/6.4 days drying, respectively. Furthermore, reduction of P2 increased progressively with the drying periods by 10.8, 13.3, 12.6 and 26.4% under continuous wetting, 3.2 days wetting/1 day drying, 3.2 days wetting/3.2 days drying and 3.2 days wetting/6.4 days drying, respectively. Reduction of protein-like fractions in soil column experiments is ascribed to breakdown of these substances into non-fluorescence structures (Xue et al., 2009). P3 was biologically reduced in soil columns exhibiting 49.3, 52.5, 51.3 and 61.9% reduction under continuous wetting, 3.2 days wetting/1 day drying, 3.2 days wetting/3.2 days drying and 3.2 days wetting/6.4 days drying, respectively. Table 5.2 shows F-EEM peaks reduction at various operating conditions.

Table 5.2 Change in fluorescence peaks intensity during application of PE at different wetting/drying cycles (media size: 0.8–1.25 mm, HLR = 1.25 m/d, temperature: 20-22°C)

Operating condition	Intensity change in fluorescence peaks (%)		
	P1 ($\lambda_{Ex/Em}$ = 240-250/425-445 nm)	P2($\lambda_{Ex/Em}$ = 325-330/424-430 nm)	P3 ($\lambda_{Ex/Em}$ = 270-280/340-350 nm)
Continuous wetting	13.0	10.8	49.3
3.2 d wetting/ 1 d drying	14.8	13.3	52.5
3.2 d wetting/ 3.2 d drying	20.3	12.6	51.3
3.2 d wetting/ 6.4 d drying	31	26.4	61.9

5.3.2.3 Nitrogen

NH_4-N reduction of 14.7±3.4, 16.8±6.5, 33.0±6.5 and 82.3±4.4% was achieved in the upper 50 cm of the column while a significant (P = 0.001) overall NH_4-N reduction of 20.1±3.7% (effluent 39.3±2.5 mg N/L), 28.7±7.8% (effluent 34.9±2.7 mg N/L), 66.9±2.2% (effluent 19.4±6.9 mg N/L) and 88.4±0.8% (effluent 4.6±0.2 mg N/L) was observed under continuous, intermittent application of PE at 3.2 days wetting/1 day drying, 3.2 days wet/3.2 days dry and 3.2 days wetting/6.4 days drying, respectively. NO_3-N concentration increased markedly in the upmost 50 cm of the column with the highest values obtained at 3.2 days wetting/6.4 days drying. NO_3-N increased from an influent value of 0.7±0.1 to 10.8±0.5 mg N/L at 50 cm, and then decreased marginally to 8.4±0.1 mg N/L at the effluent. This succession of decrease in NH_4-N and increase in NO_3-N concentrations in the upper 50 cm depth of the column was accompanied by decrease in DO from an average influent concentration of 7.3±0.1 to 0.6±0.0 mg O_2/L at 50 cm. Nevertheless, a modest DO reduction was observed at the column outlet where an average DO value of 0.4±0.0 mg O_2/L was obtained. NH_4-N was notably (P = 0.001) reduced during the longest drying cycle due to aeration of the top part of the column. Profiles of NH_4-N and NO_3-N concentrations are presented in Figure 5.5.

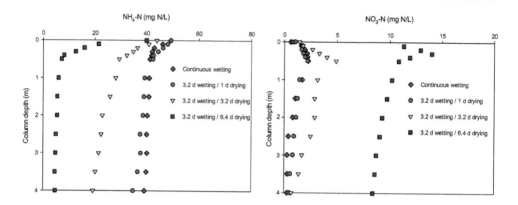

Figure 5.5 Average NH$_4$-N and NO$_3$-N concentrations along the soil column operated at continuous application and alternate wetting/drying cycles (influent: primary effluent, media size: 0.8–1.25 mm, HLR = 1.25 m/d)

5.3.2.4 E. Coli and total coliforms removal

E. coli was considerably (P = 0.004) removed at 2.5±0.2, 2.7±0.3, 3.3±0.4 and 4.2±0.0 log$_{10}$ units under continuous wetting, 3.2 days wetting/1 day drying, 3.2 days wetting/3.2 days drying and 3.2 days wetting/6.4 days drying, respectively, at the HLR of 1.25 m/d. Likewise, removal of *total coliforms* (P = 0.001) accounted to 2.5±0.2, 2.8±0.1, 3.1±0.0 and 4.1±0.0 log$_{10}$ reduction under the same operating conditions showing increasing pathogens removal with increase in drying period. Though pathogens removal increased with the length of the drying period, these results are similar to those achieved in the soil column operated at HLR 0.625 m/d suggesting that elimination of *E. coli* and *total coliforms* was predominantly achieved in the top part of the column. Figure 5.6 presents soil column removal performance of *E. coli* and *total coliforms* under different wetting/drying operating conditions.

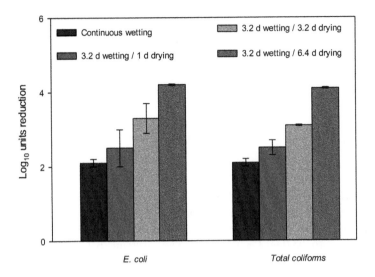

Figure 5.6 Pathogens indicators reduction during continuous and intermittent infiltration of primary effluent in soil column (media size: 0.8–1.25 mm, HLR = 1.25 m/d)

Generally, cyclic operation of infiltration basin in a MAR system helps in restoration of infiltration rates and dictates the frequency at which the basin is cleaned. While longer wetting periods could compromise the quality of SAT filtrate and result in breakthrough of ammonium (Pescod, 1992), long drying periods increase aeration of the soil column beneath the infiltration basin and improve the quality of SAT filtrate. However, longer drying cycles may have economic and operational implications since more land area will be required to divert wastewater effluent to parallel infiltration basins. This implies that more land area, distribution system and field workers will be required. For SAT systems fed with PE, cyclic operation of the basin is essential and relatively short wetting period followed by a drying period equal/twice the wetting period will effectively reduce ammonium, restore infiltration rates and allow frequent cleaning of the basin at reasonable intervals. When using SAT to reclaim wastewater effluent (i.e. PE), factors such as land availability, depth of the vadose zone, soil type, demand and market availability for the reclaimed water should be considered. In places where land areas are available with relatively fine textured soils, relatively smaller HLRs (i.e. 0.625 m/d) PE could be applied using wetting/drying ratios of 1:0.5 and 1:1. However, where relatively coarser textured soil formations and high demands for reclaimed water prevail, wetting/drying of 1:0.5 and 1:1 could be applied at higher HLRs (i.e. 1.25 m/d). Reuse applications of such water would include agricultural irrigation, landscape irrigation and industrial cooling.

5.4 CONCLUSIONS

No significant difference in suspended solids removal was found under continuous and intermittent application of PE to laboratory-based SAT system. Furthermore, SS removal was observed to be independent of the drying period as more than 70% of SS content of PE was removed in the upper 50 cm of the soil column, primarily by straining.

DOC removal by SAT during continuous and intermittent application of PE ranged from 50-60% and was not affected by change in HLR and wetting/drying cycles since most the removal occurred in the top 50 cm below the media surface.

Reduction of NH_4-N increased significantly with increase in drying period suggesting positive effect on NH_4-N reduction as compared to continuous application of PE. NH_4-N reduction at HLR of 0.625 m/d increased from 36.8±7.6% under continuous application of PE to 98.0±0.1% when 6.4 days wetting/6.4 days drying operating conditions were adopted. Nevertheless, continuous application of PE at HLR of 1.25 m/d resulted in NH_4-N reduction of 20.1±3.7% whereas cyclic wetting/drying operating conditions of 3.2 days/6.4 reduced NH_4-N by 88.4±0.8% at the same HLR. Furthermore, corresponding high NO3-N concentrations were observed in the top part of the soil column operated at HLR of 0.625 m/d compared to the one operated at HLR of 1.25 m/d when 6.4 days drying was adopted for both columns, implying that long HRT in the column operated at HLR of 0.625 m/d enabled long contact between NH_4-N contained in PE and the sorbed DO during the first wetting cycle that succeeded the drying period.

Removal of both *E. coli* and *total coliforms* increased progressively with drying period from as low as 2.5±0.2 log_{10} to as high as 4.3±0.5 log_{10} units at both HLRs suggesting that indicators pathogens removal was dependent on the length of drying period. The study demonstrated that SAT system fed with PE and operated at HLR of 0.625 m/d could be operated at wetting/drying of 1:0.5, while a wetting/drying of 1:1 is applicable at SAT systems operated at HLR of 1.25 m/d. It should be noted that longer drying periods will increase land and reclaimed water distribution system requirements.

5.5 REFERENCES

Abel, C. D. T., Sharma, S. K., Malolo, Y. N., Maeng, S. K., Kennedy, M. D. and Amy, G. L. (2012). Attenuation of Bulk Organic Matter, Nutrients (N and P), and Pathogen Indicators During Soil Passage: Effect of Temperature and Redox Conditions in Simulated Soil Aquifer Treatment (SAT). *Water, Air and Soil Pollution*, **223**, 5205-5220.

Bouwer, H. (2002). Artificial recharge of groundwater: Hydrogeology and engineering. *Hydrogeology Journal*, **10**(1), 121-142.

Fox, P., Houston, S., Westerhoff, P., Drewes, J., Nellor, M., Yanko, B., Baird, R., Rincon, M., Arnold, R. and Lansey, K. (2001). An Investigation of Soil Aquifer Treatment for Sustainable Water Reuse. *Research Project Summary of the National Center for Sustainable Water Supply (NCSWS)*, Tempe, Arizona, USA.

Greskowiak, J., Prommer, H., Massmann, G., Johnston, C., Nützmann, G. and Pekdeger, A. (2005). The impact of variably saturated conditions on hydrogeochemical changes during artificial recharge of groundwater. *Applied Geochemistry*, **20**(7), 1409-1426.

Gruenheid, S., Amy, G. and Jekel, M. (2005). Removal of bulk dissolved organic carbon (DOC) and trace organic compounds by bank filtration and artificial recharge. *Water Research,* **39**(14), 3219-3228.

Hijnen, W. A. M., Brouwer-Hanzens, A. J., Charles, K. J. and Medema, G. J. (2005). Transport of MS2 Phage, Escherichia coli, Clostridium perfringens, Cryptosporidium parvum, and Giardia intestinalis in a Gravel and a Sandy Soil. *Environmental Science and Technology,* **39**(20), 7860-7868.

Ickeson-Tal, N., Avraham, O., Sack, J. and Cikurel, H. (2003). Water reuse in Israel-the Dan Region Project: Evaluation of water quality and reliability of plant's operation. *Water Science and Technology: Water Supply,* **3**(4), 231-237.

Kopchynski, T., Fox, P., Alsmadi, B. and Berner, M. (1996). The effects of soil type and effluent pre-treatment on soil aquifer treatment. *Water Science and Technology,* **34**(11), 235-242.

Levantesi, C., La Mantia, R., Masciopinto, C., Böckelmann, U., Ayuso-Gabella, M. N., Salgot, M., Tandoi, V., Van Houtte, E., Wintgens, T. and Grohmann, E. (2010). Quantification of pathogenic microorganisms and microbial indicators in three wastewater reclamation and managed aquifer recharge facilities in Europe. *Science of the Total Environment,* **408**(21), 4923-4930.

Maeng, S. K., Sharma, S. K., Abel, C. D. T., Magic-Knezev, A., Song, K. G. and Amy, G. L. (2012). Effects of effluent organic matter characteristics on the removal of bulk organic matter and selected pharmaceutically active compounds during managed aquifer recharge: Column study. *Journal of Contaminant Hydrology,* **140-141**, 139-149.

McDowell-Boyer, L., Hunt, J. and Sitar, N. (1986). Particle transport through porous media. *Water Resources Research,* **22**(13), 1901-1921.

NRC. (2012). Water Reuse: Potential for Expanding the Nation's Water Supply Through Reuse of Municipal Wastewater, National Research Council. National Academy Press, Washington, D. C, USA.

Pescod, M. (1992). Wastewater Treatment and Use in Agriculture. Food and Agriculture Organization Irrigation and Drainage Paper 47. Rome.

Quanrud, D., Arnold, R., Lansey, K., Begay, C., Ela, W. and Gandolfi, A. (2003). Fate of effluent organic matter during soil aquifer treatment: biodegradability, chlorine reactivity and genotoxicity. *Journal of Water and Health,* **1**(1), 33-45.

Sharma, S. K., Hussen, M. and Amy, G. L. (2011). Soil aquifer treatment using advanced primary effluent. *Water Science and Technology,* **64**(3), 640-646.

van Cuyk, S., Siegrist, R., Logan, A., Masson, S., Fischer, E. and Figueroa, L. (2001). Hydraulic and purification behaviors and their interactions during wastewater treatment in soil infiltration systems. *Water Research,* **35**(4), 953-964.

Westerhoff, P. and Pinney, M. (2000). Dissolved organic carbon transformations during laboratory-scale groundwater recharge using lagoon-treated wastewater. *Waste Management,* **20**(1), 75-83.

Xue, S., Zhao, Q., Wei, L. and Ren, N. (2009). Behavior and characteristics of dissolved organic matter during column studies of soil aquifer treatment. *Water Research,* **43**(2), 499-507.

CHAPTER 6

EFFECT OF BIOLOGICAL ACTIVITY ON REMOVAL OF BULK ORGANIC MATTER, NITROGEN AND PHARMACEUTICALLY ACTIVE COMPOUNDS FROM PRIMARY EFFLUENT[4]

SUMMARY

Reduction of bulk organic matter, nitrogen and pharmaceutically active compounds from primary effluent during managed aquifer recharge was investigated using laboratory-scale batch reactors. Biologically stable batch reactors were spiked with different concentrations of sodium azide to inhibit biological activity and probe the effect of microbial activity on attenuation of various pollutants of concern. The experimental results obtained revealed that removal of dissolved organic carbon correlated with active microbial biomass. Furthermore, addition of 2 mM of sodium azide affected nitrite oxidizing bacteria leading to accumulation of nitrite-nitrogen in the reactors while an ammonium-nitrogen reduction of 95.5% was achieved. Removal efficiencies of the hydrophilic neutral compounds of phenacetin, paracetamol and caffeine were independent of the extent of the active microbial biomass present and were >90% in all reactors whereas removal of pentoxifylline was dependent on biological stability of the reactor. However, gemfibrozil, diclofenac and bezafibrate removal was >80% in batch reactors with the highest biological activity as evidenced by high concentration of adenosine triphosphate.

---

[4] Based on Abel et al. (2013). *Water, Air and Soil Pollution* **224**(7), 1-12
[4] Based on Abel et al. (2013b). Proceedings of IWA Reuse Conference. October 27-31, Windhoek, Namibia

6.1 INTRODUCTION

Presence of low concentrations of organic matter and organic matter in SAT reclaimed water is considered a major water quality and health concern, especially in SAT systems that involve indirect potable reuse of the reclaimed water (Díaz-Cruz and Barceló, 2008; Fox et al., 2001). The type and bioavailability of wastewater effluent organic matter (EfOM) affects the extent of soil biomass growth in managed aquifer recharge (MAR) systems (Rauch and Drewes, 2004) such as SAT. EfOM is measured as dissolved organic carbon (DOC) and contains low and high molecular weight substances, such as polysaccharides, proteins, aminosugars, nucleic acids, humic and fulvic acids, and cell components (Barker et al., 2000). On the other hand, the presence of nitrogen in wastewater can pose a public health hazard and affects the suitability of water for reuse (Smith, 2003). Most of the nitrogen exists in the wastewater in the form of organic-, ammonium- and nitrate-nitrogen (Sedlak, 1991). Furthermore, several pharmaceutically active compounds (PhACs) from different medicinal and prescription classes are frequently detected at influent and effluent of wastewater treatment plants (WWTPs) as well as in the receiving water downstream of the WWTPs outfalls (Heberer et al., 2002). PhACs consist of pharmaceuticals used to cure humans and animals diseases (Comerton et al., 2009). These PhACs include neutral and acidic compounds such as anelgesic/anti-inflammetory, lipid regulators and antiepileptics which have been detected at concentrations up to micrograms per liter in wastewater influent and effluents in Europe and North America (Jones et al., 2005; Metcalfe et al., 2003; Ternes, 1998). Detection of PhACs at such concentrations is ascribed to the fact that WWTPs do not provide adequate removal for PhACs (Reif et al., 2008) since these WWTPs have not been specifically designed to remove PhACs (Carballa et al., 2004) and may therefore reach water receptors and cause adverse impact on aquatic life (Miège et al., 2009). Presence of residues of organic micropollutants in the environment has been repeatedly reported to have drastic effects on aquatic ecosystems. Low sperm counts, poor sperm motility, and low milt volume observed in wild fishes downstream of WWTPs outfalls were attributed to water contamination with high concentrations of organic micropollutants (Edwards et al., 2006). Furthermore, extinction of 95% of vultures' population in India and Pakistan in 1990s was ascribed to their exposure to elevated concentrations of veterinary diclofenac (Oaks et al., 2004).

Removal of bulk organic matter, nitrogen, phosphorus, suspended solids, bacteria and viruses in soil infiltration system (i.e. SAT) is attained through sorption, chemical reaction, bio-transformation, die-off and predation (Kanarek and Michail, 1996). Although the main removal mechanism for DOC is biotic reaction, another removal mechanism that contributes to bulk organics matter removal during SAT is abiotic (Xue et al., 2009). Ammonium is biologically transformed to nitrate with nitrite as an intermediate step through nitrification (Vymazal, 2007) in the presence of a carbon source and dissolved oxygen. Besides, adsorption of ammonium to aquifer material in a SAT system is another mechanism that contributes to the overall reduction of ammonium in both unsaturated zone and the subsurface where low DO concentration prevails (Abel et al., 2013). Frequent drying of the infiltration basin allows gaseous oxygen to oxidize the sorbed ammonium.

Removal of PhACs is governed by their chemical properties, microbial activity and prevailing environmental conditions (Johnson and Sumpter, 2001). Adsorption, biodegradation and volatilization are the most important removal mechanisms in transport and fate of organic micropollutants (i.e. PhACs) in soil percolation systems (Bouwer et al., 1981). While properties of the adsorbing compound like water solubility, size, shape, configration and charge of the molcule affect its adsorption (Bedding et al., 1983), biodegradability of PhACs is affected by numerous environmental factors such as dissolved oxygen, pH, temperature and microbial activity. Hydrphobic PhACs with low water solubility easily adsorb onto solid phases. Furthermore, biodegradation of PhACs is achieved through the uptake of a growth substrate (carbon source) by microorganisms coupled with release of enzymes that facilitate the removal of PhACs.

Currently, analysis of water and wastewater to check the presence of organic micropollutants (OMPs) is not frequently performed in developing countries due to lack of tools, expertise and high investment capital requirements. However, concentrations of OMPs in water streams in the developing world are poised to worsen due to anticipated increase in the volume of wastewater generated and discharged to water streams. Besides, increase in population life expectancy coupled with availability of medicines sold without prescription at affordable prices will necessitate nalaysis and regulation of OMPs in developing countries.

The objective of this chapter was to explore the effect of biological activity on the removal of bulk organic matter, nitrogen and PhACs from primary effluent (PE) in SAT system. This study gives an insight on contributions of physical and biological mechanisms on the removal of these contaminants based on reactor's operating conditions.

6.2   MATERIALS AND METHODS

6.2.1   Chemicals

Thirteen selected pharmaceuticals from different medicinal classes such as analgesic, lipid regulators, antiepileptic agents, psychiatric drugs and vasodilators frequently detected in wastewater effluents in Europe and North America were selected. Analytical grade chemicals of the selected PhACs were purchased from Sigma-Aldrich, Germany. Physico-chemical properties of these PhACs are presented in Table 6.1. Stock solution containing a cocktail of the PhACs was prepared by dissolving 100 mg from each chemical in 100 mL ethanol giving a 1 g/L solution. Working solutions were prepared using the stock solution by spiking PE with PhACs and used as influent water for batch experiments.

Table 6.1 Physico-chemical properties of the selected PhACs

Compound	MW (g/mol)	pK$_a$	log K$_{ow}$[1]	logD[2] (pH = 8)	Charge (pH = 8)
Gemfibrozil	250.3	4.7	4.77	2.22	Ionic
Diclofenac	296.2	4.2	4.51	1.59	Ionic
Bezafibrate	361.8	3.6	4.25	0.69	Ionic
Ibuprofen	206.3	4.9	3.97	1.44	Ionic
Fenoprofen	242.3	4.5	3.9	1.11	Ionic
Naproxen	230.3	4.2	3.18	0.05	Ionic
Ketoprofen	254.3	4.5	3.12	0.41	Ionic
Clofibric acid	214.6	3.2	2.88	-1.08	Ionic
Carbamazepine	236.3	n.a.	2.45	n.a.	Neutral
Phenacetin	179.2	n.a.	1.67	n.a.	Neutral
Paracetamol	151.2	n.a.	0.27	n.a.	Neutral
Pentoxifylline	278.3	n.a.	0.29	n.a.	Neutral
Caffeine	194.2	n.a.	-0.07	n.a.	Neutral

[1] Octanol-water partition coefficient (hydrophobic: log K$_{ow}$ >2, hydrophilic: log K$_{ow}$ <2)
[2] Distribution coefficient

6.2.2 Source water characteristics

PE used in experimentation was collected from Harnaschpolder WWTP. Characteristics of the PE are presented in section 3.2.1. Furthermore, secondary effluent (SE) was collected alongside PE. Both PE and SE were then analyzed for target compounds to obtain their background concentrations in these wastewater effluents as shown in Table 6.2.

Table 6.2 Concentrations of the selected PhACs detected in PE and SE from Harnaschpolder WWTP

Compound	Therapeutic use	Primary effluent (µg/L)	Secondary effluent (µg/L)
Gemfibrozil	Lipid regulator	0.92	0.18
Diclofenac	Analgesic	0.70	0.46
Bezafibrate	Lipid regulator	0.37	0.07
Ibuprofen	Analgesic	5.00	0.10
Fenoprofen	Analgesic	0.04	0.04
Naproxen	Analgesic	4.70	0.12
Ketoprofen	Analgesic	0.11	0.04
Clofibric acid	Metabolite of lipid regulator drug	0.04	0.04
Carbamazepine	Antiepileptic, Psychiatric drug	0.56	0.77
Phenacetin	Analgesic	0.24	0.04
Paracetamol	Mild analgesic	180	0.04
Pentoxifylline	Vasodilator	0.04	0.04
Caffeine	Psychomotor stimulant	100	0.13

6.2.3 Experimental setup

To examine the effect of biological activity on the removal of bulk organic matter, nitrogen and PhACs, batch experiments were conducted in 0.5-L glass bottles containing 100 g standard silica sand with grain size ranging from 0.8 to 1.25 mm and filled with 400 mL of PE. Five sets of batch experiments were performed in duplicate for different operating conditions: (i) only PE in autoclaved batch reactors (without silica sand) as blank (PE-blank), (ii) PE added to autoclaved silica sand at 120°C (PE-control), (iii) PE in a biologically stable batch reactor (PE-bioactive), (iv) PE in a biologically stable batch reactors spiked with 2 mM sodium azide (NaN_3) (PE+2 mM-NaN_3), and, (v) PE in a biologically stable batch reactors spiked with 20 mM of NaN_3 (PE+20 mM-NaN_3). All glass bottles used in the experiments were autoclaved before the addition of silica sand and PE at 120°C for 30 minutes. Additionally, the silica sand used in the experiments was autoclaved under similar conditions, air-dried at room temperature, washed with acidified water (2 M HCl) to get rid of any functional groups, soaked in demineralized water for 24 hours, air-dried and finally, filled into the reactors. While the silica sand was added to PE+2 mM-NaN_3, PE+20 mM-NaN_3 and PE-bioactive were autoclaved once and used throughout the experimental time, while new autoclaved silica sand was added to PE-blank and PE-control were autoclaved (every five days). Additionally, the silica sand used in the experiments was autoclaved under similar conditions, air-dried at room temperature, washed with acidified water (2 M HCl) to get rid of any functional groups, soaked in demineralized water for 24 hours, air-dried and finally, filled into the reactors. All the reactors were placed on a horizontal reciprocal shaker operated at a shaking speed of 100 rpm, and samples from feed and effluent water were collected before and after hydraulic residence time (HRT) of five days. PE-bioactive, PE+2 mM-NaN_3 and PE+20 mM-NaN_3 reactors were aerobically ripened at room temperature for a period of 120 days with respect to DOC removal until a difference of ±1% was attained between each three successive DOC measurements. The reactors were then fed with PE and PE spiked with respective NaN_3 concentrations for another 120 days to suppress microbial growth after ripening of the reactors to balance the role of biological activity against abiotic mechanisms in removal of bulk organic matter, nitrogen and PhACs. Out of the thirteen PhACs studied, paracetamol and caffeine were detected in PE at relatively high concentrations, whereas concentrations of the remaining PhACs ranged between 0.04 µg/L and 5 µg/L. The PE fed to batch reactors was spiked with a concentration of 5 µg/L for each PhAC based on the occurrence of the target compounds in influent and effluent of WWTPs and aquatic environment in Europe at concentrations up to µg/L level. This uniform concentration for all the PhACs was adopted to maintain their concentrations at µg/L level and account for any change in the quality of the PE sampled from the WWTP. Concentrations of PhACs in influent water after spiking PE with a cocktail of the target compounds ranged from 2.4±0.0 µg/L (phenacetin) to 12.5±0.7 µg/L (ibuprofen) while concentrations of caffeine and paracetamol were 110.0±0.0 and 177.0±11.3 µg/L, respectively. Reduction in the initial concentration of phenacetin in the spiked PE (<5 µg/L) may be ascribed to the presence of active extracellular enzymes in PE induced by low biological activity in primary sedimentation tank and DOC biodegradation. However, paracetamol and caffeine were detected at concentrations higher than100 µg/L in PE background. Measurements of DOC, ammonium-nitrogen (NH_4-N), nitrite-nitrogen (NO_2-N) and nitrate-nitrogen (NO_3-N) were conducted after spiking the reactors with NaN_3.

6.2.4 Analytical methods

DOC concentrations and UV absorbance at 254 nm wavelength (UVA_{254}) of all pre-filtered samples (n=18) collected from the WWTP and laboratory-scale set-ups were measured within three days using total organic carbon (TOC) analyzer and UV-VIS spectrophotometer as detailed in section 3.2.4. Specific ultraviolet absorbance (SUVA) was calculated using UV values and their corresponding DOC measurements. Organic matter content of silica sand was measured for clean sand and after five days in PE-control reactors to investigate change in soil organic matter. Silica sand samples (5 g) were weighed into 50 mL Milli-Q water spiked with 10 mL of HNO_3 and digested at $105\pm5°C$ for 30 minutes. The samples were centrifuged at 3000 rpm for 20 minutes, and TOC of the supernatant solution was measured using TOC analyzer.

Samples from various batch reactors were diluted with Milli-Q water (Advantage A10; Millipore) with respect to their DOC to obtain 1 mg/L DOC concentration without any pH adjustment. Fluorescence excitation-emission (F-EEM) spectra were then obtained through collection of a series of emission spectra at different excitation wavelengths using a FluoroMax-3 spectrofluorometer (HORIBA Jobin Yvon, Edison, NJ, USA) and MATLAB (version 7.9, R2009b) used to illustrate organic matter fractions of humic-, fulvic- and protein-like as presented in section 3.2.4. Peak 1 (P1) was assigned to (primary) humic-like, peak 2 (P2) to (secondary humic) fulvic-like while peak 3 (P3) to protein-like.

Chemical reagents used to determine ammonium as nitrogen (NH_4-N), nitrite as nitrogen (NO_2-N) and nitrate as nitrogen (NO_3-N) were of analytical grade and were purchased from Merck KGaA, Germany and J.T. BAKER, Netherlands. These nitrogenous inorganics were determined using colorimetric automated techniques with a spectrophotometer according to Eaton et al. (2005). Nine duplicate of samples (n=18) were analyzed for NH_4-N, NO_3-N, and NO_2-N.

Adenosine triphosphate (ATP) has been used to correlate active biomass present in batch reactors with biodegradable organic matter in different waters (Maeng et al., 2011). Dehydrogenase activity, phospholipid extraction and substrate induced respiration have been used to estimate how biomass associated with soil is related to dissolve organic matter during soil passage (Rauch and Drewes, 2005). Recently, ATP has been used to investigate total active microbial biomass (AMB) associated with sand in different managed aquifer systems (MAR) including bank filtration and artificial recharge and recovery (Maeng et al., 2008; Maeng et al., 2012). Therefore, to quantify AMB on the media used in various batch reactors, ATP was measured using 2.0-2.5 g of well mixed wet sand from PE-control, PE+2 mM-NaN_3, PE+20 mM-NaN_3 and PE-bioactive. The sand was suspended in a 50 mL autoclaved tap water prior to sonication phase aiming at detaching the biomass from the media (solid phase) to liquid phase (water). Additionally, ATP of PE-blank was measured using PE samples from the reactors every five days.

Measurements of PhACs (n=2) of interest were carried out according to the method detailed in Sacher et al. (2008). Enrichment of the selected PhACs was carried out using BakerbonTM SPE styrene-divinylbenzene (SDB-1, 200 mg) sorbent (Mallinckrodt Baker, Deventer, Netherlands) following solid phase extraction (SPE) using Caliper Life Science GmbH (Rüsselsheim, Germany) workstation. Measurements of PhACs were performed on high-performance liquid chromatography-electrospray ionization tandem mass spectrometry analyses (HPLC-ESI-MS-MS) using a HPLC system 1100, Series II from Agilent Technologies (Waldbronn, Germany), furnished with an API 2000 triple quadrupole mass spectrometer from PE Sciex (Langen, Germany).

6.3   RESULTS AND DISCUSSION

6.3.1   AMB and bulk organic matter

ATP concentrations of 0.011 ± 0.003, 0.514 ± 0.119, 1.243 ± 0.203, 1.828 ± 0.149 and 4.013 ± 0.557 µg ATP/cm^3 were measured in PE-blank, PE-control, PE+20 mM-NaN$_3$, PE+2 mM-NaN$_3$ and PE-bioactive reactors, respectively. These ATP concentrations corresponded with DOC removals of 14.3 ± 5.0, 30.4 ± 5.0, 45.0 ± 5.6, 64.6 ± 2.6 and $75.3\pm2.4\%$ under identical conditions. Table 6.3 shows DOC concentrations and removal with respect to biological activity in various reactors.

Table 6.3 DOC removal as a function of microbial activity

Reactor	Influent DOC (mg/L)	Effluent DOC (mg/L)	DOC removal (%)	ATP (µg ATP/cm^3)
PE-blank	44.9 ± 4.4	38.3 ± 2.4	14.3 ± 5.0	0.011 ± 0.003
PE - control	44.9 ± 4.4	31.2 ± 3.2	30.4 ± 5.0	0.514 ± 0.110
PE + 20 mM-NaN$_3$	44.9 ± 4.4	24.7 ± 3.3	45.0 ± 5.6	1.243 ± 0.203
PE + 2 mM-NaN$_3$	44.9 ± 4.4	15.8 ± 1.5	64.6 ± 2.6	1.828 ± 0.149
PE + bioactive	44.9 ± 4.4	11.0 ± 0.9	75.3 ± 2.4	4.013 ± 0.557

These results demonstrate that PE-blank reactor had the lowest biological activity followed by PE-control. Furthermore, addition of sodium azide caused biological activity reduction in PE+20 mM-NaN$_3$ and PE+2 mM-NaN$_3$ batch reactors, whereas PE-bioactive reactor revealed the highest biological activity. Correlation between DOC removal and ATP concentration in different reactors is presented in Figure 6.1.

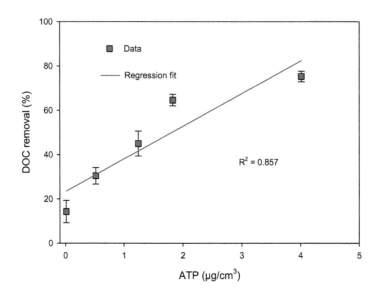

Figure 6.1 Effect of microbial activity on DOC removal

Removal of DOC in PE-blank batch reactor might be due to combination of DOC sorption to reactor surface and presumably DOC uptake by PE biota in the freshly formed biofilms on that surface. On the other hand, DOC removal in PE-control reactor may be ascribed to combination of sorption to the sand media, reactor surface and biological removal by naturally occurring bacteria in biofilms associated with the media as evidenced by decrease in SUVA and the presence of 0.514 ± 0.119 μg ATP/cm^3 compared to 0.011 ± 0.003 ATP/cm^3 in PE-blank reactor. Improved DOC removal in PE-control reactor ($30.4\pm3.7\%$) as compared to PE-blank reactor ($14.3\pm5.0\%$) accounted for a $16.1\pm4.7\%$ difference which is presumably ascribed to sorption and biodegradation of DOC induced by the relatively high microbial activity in PE-control. Furthermore, DOC removal in PE-bioactive reactor was achieved through biodegradation and sorption of DOC onto the biofilms around silica sand. Part of the organic matter sorbed onto soil is biologically degraded (Idelovitch et al., 2003). However, addition of 2 and 20 mM NaN_3 solutions affected the biological activity of the biologically active reactors and resulted in decrease in DOC removal by 10 and 30% in PE+2 mM NaN_3 and PE+20 mM NaN_3 reactors, respectively. These results suggest that NaN_3 did not completely eliminate biological activity as evidenced by notable DOC removal in PE+2 mM NaN_3 and PE+20 mM NaN_3 reactors. The observed reduction of DOC removal is due to addition of different concentrations of NaN_3 which eliminated part of the AMB as suggested by less ATP concentrations in these reactors compared to PE-bioactive. This is in disagreement with Cha et al. (2004) who postulated that addition of 2 mM NaN_3 inhibited biological activity in batch experiments.

SUVA values decreased from 2.59 ± 0.23 to 2.36 ± 0.24 and 2.06 ± 0.18 L/mg. m in PE-blank and PE-control reactors, respectively. Conversely, effluent samples from PE-bioactive, and PE+2 mM-NaN$_3$ reactors exhibited increase in SUVA from 2.59±0.23 to 3.46±0.48, and 3.28±0.34 L/mg. m, respectively. SUVA depicts relative aromaticity and contribution of DOC aromatic structures (Quanrud et al., 1996). Reduction in SUVA values of effluent samples from laboratory-scale batch reactors suggests preferential removal of higher molecular weight, hydrophobic and aromatic DOC constituents through sorption (McCarthy et al., 1993; Westerhoff and Pinney, 2000). Decrease in SUVA of PE-blank and PE-control is attributed to removal of aromatic organic matter through sorption. Such removal pattern is reported in the literature (Gruenheid et al., 2005; Quanrud et al., 1996). On the other hand, biodegradation of lower molecular weight DOC substances and release of high molecular weight soluble microbial products (SMPs) contribute to increase in SUVA (McCarthy et al., 1993). SUVA markedly increased from 2.59±0.23 L/mg. m in the influent to 3.46±0.48 and 3.28±0.34 L/mg. m in PE-bioactive and PE+2 mM-NaN$_3$, respectively, suggesting preferential removal of aliphatic (easily biodegradable) fractions of DOC.

Reduction of fluorescence intensities was observed for all F-EEM spectra peaks (P1, P2 and P3) of effluent samples from blank, control, and azide spiked batch reactors. Change in fluorescence intensity for each of the peaks identified was calculated as the difference between fluorescence intensity of influent peak and the corresponding peak of effluent sample from each batch reactor. Humic (humic and fulvic) substances are negatively charged (Laangmark et al., 2004) and could be therefore, removed by sorption to soil (Quanrud et al., 1996). The relatively low reduction of P1 and P2 intensities in most reactors could be attributed to properties of silica sand which bears negative charges. However, high reduction in P3 is attributed to break down of protein-like substances into non-fluorescent substances. Such removal was observed in soil column experiments simulating SAT by Xue et al. (2009). High biological activity in PE-bioactive resulted in P3 reduction of 62.1% while PE+2 mM-NaN$_3$ and PE+20 mM-NaN$_3$ exhibited drop in P3 reduction by 14 and 24.2%, respectively. This decrease in the reduction of P3 due to addition of NaN$_3$ is consistent with reduction in DOC removal. 3D plots and wavelength ranges ($\lambda_{ex/em}$) at which the peaks were observed for influent, PE-blank, PE-control, PE+20 mM-NaN$_3$, PE+2 mM-NaN$_3$ and PE-bioactive samples are presented in Figure 6.2 and Table 6.4.

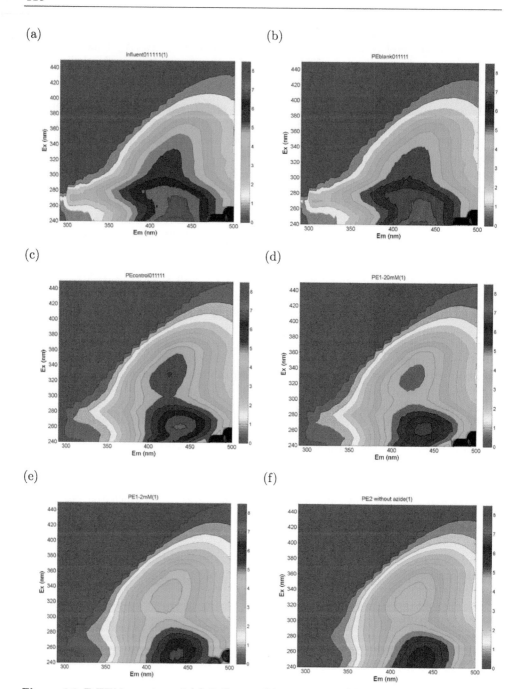

Figure 6.2 F-EEM spectra of (a) Influent, (b) PE-blank, (c) PE-control, (d) PE+20 mM-NaN$_3$, (e) PE+2 mM-NaN$_3$ and (f) PE-bioactive (media size: 0.8–1.25 mm, HRT: 5 days, temperature: 20-22°C)

Table 6.4 Change in fluorescence peaks intensities in aerobic batch reactors operated under different biological activity (Influent: PE, media size: 0.8–1.25 mm, HLR = 0.625 m/d, temperature: 20-22°C)

Sample	Fluorescence Peaks		
	P1 (λex/em = 240-260/434-438 nm)	P2 (λex/em = 330-340/420-430 nm)	P3 (λex/em = 270-280/350-360 nm)
Influent	4.94	3.75	3.22
PE-blank	4.90	3.64	2.70
Change (%)	0.8 (-)	2.93 (-)	16.1 (-)
PE-control	4.17	3.08	2.20
Change (%)	15.6 (-)	17.9 (-)	31.7 (-)
(PE+20 mM-NaN$_3$)	4.60	3.72	2.00
Change (%)	6.9 (-)	0.8 (-)	37.9 (-)
(PE+2 mM-NaN$_3$)	4.33	3.53	1.67
Change (%)	12.3 (-)	5.9 (-)	48.1 (-)
PE-bioactive	3.88	3.22	1.22
Change (%)	21.5 (-)	14.1 (-)	62.1 (-)

6.3.2 Nitrogen

In order to probe the contribution of physical removal mechanism in NH_4-N reduction, NH_4-N content of influent and effluent samples was quantified. It was noted that influent NH_4-N concentration of 47.1±3.4 mg N/L slightly decreased to 46.6±3.4 and 42.0±3.1 mg N/L accounting to 0.6±0.4 and 10.4±2.5% in PE-blank and PE-control, respectively. While addition of 2 mM NaN_3 did not result in suppression of NH_4-N oxidizing bacteria as evidenced by 95.5±0.5% (effluent 2.1±0.2 mg N/L) NH_4-N reduction, batch reactors spiked with a 20 mM NaN_3 exhibited NH_4-N reduction of 29.5±5.8% (effluent 33.1±3.7 mg N/L). However, PE-bioactive revealed 99.9±0.0% (effluent 0.3±0.03 mg N/L) NH_4-N reduction. Consequently, NO3-N concentration changed from 0.17±0.02 mg N/L in the influent to 0.18±0.01, 0.18±0.02, 0.3±0.03, 3.75±0.61 and 36.95±1.38 mg N/L in PE-blank, PE-control, PE+20 mM-NaN_3, PE+2 mM-NaN_3 and PE-bioactive, respectively. Figure 6.3 presents change in NH_4-N concentration of PE in batch reactors operated under different biological activity.

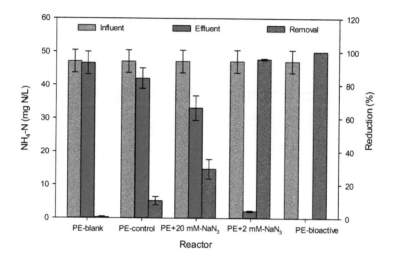

Figure 6.3 Change in NH$_4$-N concentrations of PE in aerobic batch experiments (media size, 0.8–1.25 mm, HRT, 5 days)

As shown in Figure 6.3, PE-blank and PE-control exhibited the lowest NH$_4$-N reduction. NH$_4$-N reduction of 10.4±2.5% in PE-control is presumably due to its sorption onto the media. Nonetheless, attenuation of NH$_4$-N in PE+20 mM-NaN$_3$, PE+2 mM-NaN$_3$ and PE-bioactive could be ascribed to combine effect of nitrification and sorption with the highest reduction attained in PE-bioactive. Nitrification is a two-step autotrophic oxidation process in which NH$_4^+$ oxidizing bacteria catalyzes conversion of NH$_4^+$ to NO$_2^-$ followed by oxidation of NO$_2^-$ to NO$_3^-$ in the presence of NO$_2^-$ oxidizing bacteria. NH$_4$-N reduction of 95.5±0.5% in PE+2 mM-NaN$_3$ indicated that nitrification process remained functional in these reactors. Nevertheless, NO$_2^-$ increased from 0.04±0.01 mg N/L in the influent to 27.17±2.58 mg N/L in effluent samples; coupled with a relatively low NO$_3$-N concentration of 3.75±0.61 mg N/L. Low NO$_3$-N generation suggests that NO$_2^-$ oxidizing bacteria were negatively affected by addition of 2 mM NaN$_3$. Furthermore, addition of 20 mM-NaN$_3$ caused NH$_4$-N reduction to decrease significantly from 99.9±0.0% in PE-bioactive to 29.5±5.8% revealing the severe effect of the biocide on both NH$_4^+$ and NO$_2^-$ oxidizing bacteria as suggested by slight increase in NO$_2$-N concentration from 0.04±0.01 mg N/L to 0.33±0.09 mg N/L and effluent NO$_3$-N of 0.3±0.03 mg N/L.

6.3.3 Removal of selected PhACs

Of the 13 PhACs studied, removal of ionic compounds (gemfibrozil, diclofenac, bezafibrate, ibuprofen and naproxen) in PE-blank batch reactors was below 5%. However, the fate of some of these compounds (such as bezafibrate and naproxen) remained unchanged in PE-control while ibuprofen was significantly removed by 75.8±8.2%. Removals of other ionic compounds, such as ketoprofen and clofibric acid and neutral compound carbamazepine removals in PE-blank and PE-control ranged from 10-30%. Furthermore, removal efficiencies of the hydrophilic neutral compounds phenacetin, paracetamol and caffeine were greater than 95%. The neutral vasodilator pentoxifylline was removed by 5.7±1.5% in PE-blank and PE-control. Figure 6.4 illustrates removals of the selected compounds in blank and control samples from batch experiments.

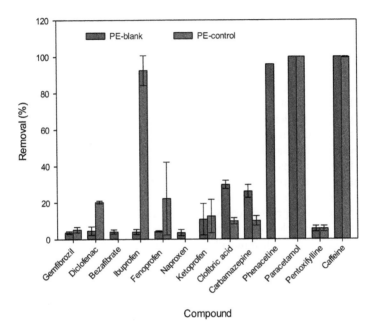

Compound

Figure 6.4 Effect of abiotic conditions on removal of PhACs from PE in batch reactors (media size, 0.8–1.25 mm, HRT 5 days)

Removal of hydrophobic ionic compounds gemfibrozil, diclofenac, bezafibrate, ibuprofen and naproxen in PE-blank might be due to sorption to the surface of the reactors. Sorption of neutral anelgesic drug ibuprofen to particles combined with particulate organic carbon enhances its removal by sedimentation (Tixier et al., 2003). Improved removal of ibuprofen in PE-control as compared to PE-blank could be attributed to sorption to organic carbon since total organic carbon (TOC) of silica sand increased from 84.5±5.9 to 122.5±3.2 μg TOC/g after HRT of five days. The hydrophilic neutral compounds phenacetin, paracetamol and caffeine were considerably removed presumably through co-metabolism as relatively low concentrations of PhACs could not support microorganism growth as primary carbon source substrate compared to other contaminants. Biological removal of PhACs in PE-blank and PE-control took place as wastewater biota acclimatized to PhACs which were detected at microgram concentrations in PE. The low removal of pentoxifylline in PE-control is attributed to insufficient ripening of the reactors as this compound is well removed in biostable systems. This is consistent with the findings of Maeng et al. (2011) in which pentoxifylline removal was found to be dependent on ripening duration of batch reactors.

Removal of most of the compounds of interest was >90% in PE-bioactive reactor and 2 mM-NaN$_3$ except for gemfibrozil, diclofenac, colfibric acid and carbamazepine. Addition of 20 mM NaN$_3$ to PE affected removal of some compounds. Gemfibrozil, bezafibrate and naproxen remained unchanged in PE+20 mM-NaN$_3$ reactor after HRT of 5 days. On the other hand, removal of diclofenac, fenoprofen and ketoprofen ranged from 12.5$\pm$2.5 to 42.3$\pm$2.0%. Ibuprofen was not affected by addition of NaN3 and was removed at rates >95% implying removal through sorption. Additionally, gemfibrozil, diclofenac and bezafibrate removals ranged from 2.6 to 40.4%. Removal efficiencies of hydrophilic neutral compounds phenacetin, paracetamol and caffeine were greater than 95 % in most reactors. Since hydrophilic compounds are removed through co-metabilism by microorganisms following degradation of a growth substrate, high removal of these compounds in all reactors suggests that even low biological activity (as evidenced by low DOC removal and low ATP in PE-blank and PE-control) was sufficient to support their removal. Additionally, detection of paracetamol and caffeine at high concentrations (>100 µg/L) in PE suggests that wastewater microorganisms got acclimatized to the high concentrations and remove these compounds with time (5 days). Clofibric acid and carbamazepine were comparatively removed in all the reactors with the former better removed in PE-bioactive. However, an improved removal of clofibric acid of 70.1 $\pm$ 2.3% was achieved in PE-bioactive. Since clofibric acid is better removed under anoxic operating conditions, this relatively high removal could be ascribed to formation of local anoxic anoxic zones in biofilms around the media. Although removal of gemfibrozil, diclofenac, bezafibrate, fenoprofen, naproxen and ketoprofen was relatively low under abiotic conditions (PE-blank and PE-control), improved removal was achieved under biotic conditions (PE-bioactive) presumably through biosorption in the biofilms associated with the sand media. Removal of the selected compounds in bioactive and azide spiked batch reactors is presented in Figure 6.5, whereas a summary of average PhACs concentrations measured in the influent and effluent of various batch reactors is shown in Table 6.5.

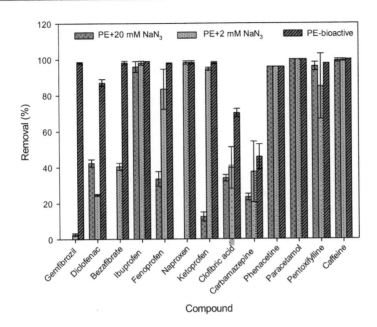

Figure 6.5 Effect of microbial activity on removal of PhACs from PE in batch reactors (media size, 0.8–1.25 mm, HRT 5 days)

Table 6.5 Summary of average PhACs in influent and effluent samples of batch experiments

Compound	PE	PE-blank	PE-control	PE+20 mM-NaN$_3$	PE+2 mM-NaN$_3$	PE-bioactive
	Inf. (µg/L)	Eff. (µg/L)	Eff. (µg/L)	Eff. (µg/L)	Eff. (µg/L)	Eff. (µg/L)
Gemfibrozil	5.6±1.1	5.4±1.1	5.3±1.0	5.9±1.1	5.5±1.1	0.1±0.0
Diclofenac	5.7±1.3	5.5±1.3	4.6±1.1	3.3±0.8	4.3±1.0	0.7±0.0
Bezafibrate	6.6±3.5	6.4±3.5	7.0±3.6	6.9±3.6	3.9±2.0	0.1±0.0
Ibuprofen	12.5±0.7	12.0±0.8	2.9±3.8	0.5±0.2	0.3±0.1	0.1±0.0
Fenoprofen	4.7±0.4	4.5±0.4	3.7±1.2	3.1±0.4	0.8±0.2	0.1±0.0
Naproxen	6.4±3.1	6.2±3.1	7.4±3.2	9.0±0.5	0.1±0.0	0.1±0.0
Ketoprofen	5.4±1.6	4.7±1.0	4.8±1.9	4.7±1.6	0.3±0.0	0.1±0.0
Clofibric acid	5.0±0.6	3.5±0.3	4.5±0.4	3.3±0.3	3.1±0.9	1.5±0.3
Carbamazepine	6.6±1.0	4.9±0.5	6.0±1.1	5.1±0.6	4.1±0.5	3.6±0.1
Phenacetin	2.4±0.0	0.1±0.0	0.1±0.0	0.1±0.0	0.1±0.0	0.1±0.0
Paracetamol	177.0±11.3	0.1±0.0	0.1±0.0	0.1±0.0	0.1±0.0	0.1±0.0
Pentoxifylline	4.4±0.1	4.1±0.0	4.1±0.0	0.2±0.1	0.7±0.3	0.1±0.0
Caffeine	110.0±0.0	0.1±0.0	0.3±0.1	0.8±1.0	0.6±0.2	4.2 ±0.1

These results suggest that ripening of batch reactors and consequently SAT system plays a pivotal role in the removal of bulk organic matter, nitrogen and most PhACs of interest. While a ripening time of 5 days resulted in DOC removal of 30.4±5.0

mg/L, a reactor that was operated for 240 days was capable of removing 75.3±2.4% of DOC content of PE. Similar trends were observed for NH_4-N where almost complete reduction of NH_4-N was achieved in the reactors ripened for 240 days. It was also noticed that the reactors ripened for 240 days with high biological activity improved the removal of gemfibrozil, diclofenac and bezafibrate from <20% in reactors ripened for 5 days to >90%. Phenacetin, paracetamol, ibuprofen and caffeine were easily removed under various operating conditions. This implies that while new SAT systems would easily remove phenacetin, paracetamol and caffeine even during the first flooding after drying or scrapping of SAT infiltration basin, substantial removal of gemfibrozil, diclofenac, pentoxifylline and bezafibrate from PE would require high biological activity after long ripening periods (i.e. 240 days). The results also show that fully ripened SAT system (i.e. >240 days) can substantially (70.1±2.3%) remove clofibric acid.

6.4 CONCLUSIONS

The investigation illustrated that DOC removal from PE was largely influenced by the extent of microbial activity in the reactor as evidenced by ATP concentration. It was observed that the lowest DOC removal was achieved in PE-blank, whereas the highest DOC removal occurred in PE-bioactive suggesting positive correlation between AMB and DOC removal.

Low NH_4-N reduction (10.4±2.5%) was observed under relatively low biological activity in PE-control and was predominantly through sorption. However, batch reactors spiked with 2 mM NaN_3 exhibited average NH_4-N reduction of 95.5±0.5% which in turn led to modest increase in NO_3-N from 0.17±0.02 to 3.75±0.61 mg N/L. On the other hand, nitrite increased from 0.04±0.01 mg N/L to 27.17±2.58 mg N/L suggesting that nitrite oxidizing bacteria were affected by addition of 2 mM NaN_3 more than ammonium oxidizing bacteria. Furthermore, addition of 20 mM NaN_3 ceased nitrification process in the reactors.

Removal of hydrophilic neutral compounds phenacetin, paracetamol and caffeine was >95% under most operating conditions, whereas pentoxifylline was poorly removed (5.7%) in the unripened reactors (PE-blank and PE-control). However, removal of the pentoxifylline compound was >85% in PE+2 mM-NaN_3, PE+20 mM- NaN_3 and PE-bioactive due to the ample ripening period (120 days)and consequently relative biostable conditions in these reactors.

Removal of antiepileptic drug carbamazepine was below 50% in PE+2 mM-NaN_3, PE+20 mM-NaN_3 and PE-bioactive reactors. Similar persistent behavior was observed for clofibric acid except that 70% removal was attained in PE-bioactive. Except for ibuprofen which was removed by 75.8% in PE-control presumably through sorption, average removal of hydrophobic compounds (gemfibrozil, diclofenac, bezafibrate, ibuprofen, naproxen and ketoprofen) of 10% was achieved under low biological activity (PE-control). However, removal of these compounds increased notably to >90% in batch reactors with high biological activity (PE-bioactive).

The obtained results imply that while new SAT systems would easily remove phenacetin, paracetamol and caffeine even during the first flooding after drying or scrapping of SAT infiltration basin, substantial removal of gemfibrozil, diclofenac, pentoxifylline and bezafibrate from PE would require high biological activity after long ripening periods (i.e. 240 days). The results also show that fully ripened SAT system (i.e. >240 days) can substantially (70.1±2.3%) remove clofibric acid.

6.5   REFERENCES

Abel, C. D. T., Sharma, S. K., Maeng, S. K., Magic-Knezev, A., Kennedy, M. D. and Amy, G. L. (2013). Fate of Bulk Organic Matter, Nitrogen, and Pharmaceutically Active Compounds in Batch Experiments Simulating Soil Aquifer Treatment (SAT) Using Primary Effluent. *Water, Air and Soil Pollution,* **224**(7), 1-12.

Barker, D., Salvi, S., Langenhoff, A. and Stuckey, D. (2000). Soluble microbial products in ABR treating low-strength wastewater. *Journal of Environmental Engineering,* **126**(3), 239-249.

Bedding, N., McIntyre, A., Perry, R. and Lester, J. (1983). Organic contaminants in the aquatic environment II. Behaviour and fate in the hydrological cycle. *The Science of The Total Environment,* **26**(3), 255-312.

Bouwer, E., McCarty, P. and Lance, J. (1981). Trace organic behavior in soil columns during rapid infiltration of secondary wastewater. *Water Research,* **15**(1), 151-159.

Carballa, M., Omil, F., Lema, J. M., Llompart, M., García-Jares, C., Rodríguez, I., Gómez, M. and Ternes, T. (2004). Behavior of pharmaceuticals, cosmetics and hormones in a sewage treatment plant. *Water Research,* **38**(12), 2918-2926.

Cha, W., Fox, P., Mir, F. and Choi, H. (2004). Characteristics of biotic and abiotic removals of dissolved organic carbon in wastewater effluents using soil batch reactors. *Water Environment Research,* **76**(2), 130-136.

Comerton, A. M., Andrews, R. C. and Bagley, D. M. (2009). Practical overview of analytical methods for endocrine-disrupting compounds, pharmaceuticals and personal care products in water and wastewater. *Philosophical Transactions of the Royal Society A: Mathematical, Physical and Engineering Sciences,* **367**(1904), 3923-3939.

Díaz-Cruz, M. and Barceló, D. (2008). Trace organic chemicals contamination in ground water recharge. *Chemosphere,* **72**(3), 333-342.

Eaton, A. D., Clesceri, L. S., Rice, E. W. and Greenberg, A. E. (2005). *Standard Methods for the Examination of Water and Wastewater. 21th ed.* American Public Health Association, American Water Works Association, and Water Environment Federation, Washington, DC, USA.

Edwards, T. M., Moore, B. C. and Guillette, L. J. (2006). Reproductive dysgenesis in wildlife: a comparative view. *International Journal of Andrology,* **29**(1), 109-121.

Fox, P., Houston, S. and Westerhoff, P. (2001). *Soil Aquifer Treatment for Sustainable Water Reuse.* American Water Works Association, Denver, Clorado, USA.

Gruenheid, S., Amy, G. and Jekel, M. (2005). Removal of bulk dissolved organic carbon (DOC) and trace organic compounds by bank filtration and artificial recharge. *Water Research,* **39**(14), 3219-3228.

Heberer, T., Reddersen, K. and Mechlinski, A. (2002). From municipal sewage to drinking water: fate and removal of pharmaceutical residues in the aquatic environment in urban areas. *Water Science and Technology,* **46**(3), 81-88.

Idelovitch, E., Icekson-Tal, N., Avraham, O. and Michail, M. (2003). The long-term performance of Soil Aquifer Treatment(SAT) for effluent reuse. *Water Science and Technology: Water Supply,* **3**(4), 239-246.

Johnson, A. C. and Sumpter, J. P. (2001). Removal of endocrine-disrupting chemicals in activated sludge treatment works. *Environmental Science and Technology,* **35**(24), 4697-4703.

Jones, O., Voulvoulis, N. and Lester, J. (2005). Human pharmaceuticals in wastewater treatment processes. *Critical Reviews in Environmental Science and Technology,* **35**(4), 401-427.

Kanarek, A. and Michail, M. (1996). Groundwater recharge with municipal effluent: Dan region reclamation project, Israel. *Water Science and Technology,* **34**(11), 227-233.

Laangmark, J., Storey, M., Ashbolt, N. and Stenstroem, T. (2004). Artificial groundwater treatment: biofilm activity and organic carbon removal performance. *Water Research,* **38**(3), 740-748.

Maeng, S., Sharma, S., Magic-Knezev, A. and Amy, G. (2008). Fate of effluent organic matter (EfOM) and natural organic matter (NOM) through riverbank filtration. *Water Science and Technology,* **57**(12), 1999.

Maeng, S. K., Sharma, S. K., Abel, C. D. T., Magic-Knezev, A. and Amy, G. L. (2011). Role of biodegradation in the removal of pharmaceutically active compounds with different bulk organic matter characteristics through managed aquifer recharge: Batch and column studies. *Water Research,* **45**(16), 4722-4736.

Maeng, S. K., Sharma, S. K., Abel, C. D. T., Magic-Knezev, A., Song, K. G. and Amy, G. L. (2012). Effects of effluent organic matter characteristics on the removal of bulk organic matter and selected pharmaceutically active compounds during managed aquifer recharge: Column study. *Journal of Contaminant Hydrology,* **140-141**, 139-149.

McCarthy, J. F., Williams, T. M., Liang, L., Jardine, P. M., Jolley, L. W., Taylor, D. L., Palumbo, A. V. and Cooper, L. W. (1993). Mobility of natural organic matter in a study aquifer. *Environmental Science and Technology,* **27**(4), 667-676.

Metcalfe, C. D., Koenig, B. G., Bennie, D. T., Servos, M., Ternes, T. A. and Hirsch, R. (2003). Occurrence of neutral and acidic drugs in the effluents of Canadian sewage treatment plants. *Environmental Toxicology and Chemistry*, **22**(12), 2872-2880.

Miège, C., Choubert, J. M., Ribeiro, L., Eusèbe, M. and Coquery, M. (2009). Fate of pharmaceuticals and personal care products in wastewater treatment plants - Conception of a database and first results. *Environmental Pollution*, **157**(5), 1721-1726.

Oaks, J. L., Gilbert, M., Virani, M. Z., Watson, R. T., Meteyer, C. U., Rideout, B. A., Shivaprasad, H., Ahmed, S., Chaudhry, M. J. I. and Arshad, M. (2004). Diclofenac residues as the cause of vulture population decline in Pakistan. *Nature*, **427**(6975), 630-633.

Quanrud, D., Arnold, R., Wilson, L. and Conklin, M. (1996). Effect of soil type on water quality improvement during soil aquifer treatment. *Water Science and Technology*, **33**(10), 419-432.

Quanrud, D., Arnold, R., Wilson, L., Gordon, H., Graham, D. and Amy, G. (1996). Fate of organics during column studies of soil aquifer treatment. *Journal of Environmental Engineering*, **133**(4), 314-321.

Rauch, T. and Drewes, J. (2004). Assessing the removal potential of soil-aquifer treatment systems for bulk organic matter. *Water Science and Technology*, **50**(2), 245-253.

Rauch, T. and Drewes, J. (2005). Quantifying biological organic carbon removal in groundwater recharge systems. *Journal of Environmental Engineering*, **131**(6), 909-923.

Reif, R., Suárez, S., Omil, F. and Lema, J. (2008). Fate of pharmaceuticals and cosmetic ingredients during the operation of a MBR treating sewage. *Desalination*, **221**(1-3), 511-517.

Sacher, F., Ehmann, M., Gabriel, S., Graf, C. and Brauch, H. (2008). Pharmaceutical residues in the river Rhine—results of a one-decade monitoring programme. *Journal of Environmental Monitoring*, **10**(5), 664-670.

Sedlak, R. (1991). *Phosphorus and Nitrogen Removal from Municipal Wastewater: Principles and Practice*. CRC Press LLC, Boca Raton, Florida, USA.

Smith, V. H. (2003). Eutrophication of freshwater and coastal marine ecosystems a global problem. *Environmental Science and Pollution Research*, **10**(2), 126-139.

Ternes, T. (1998). Occurrence of drugs in German sewage treatment plants and rivers. *Water Research*, **32**(11), 3245-3260.

Tixier, C., Singer, H. P., Oellers, S. and Müller, S. R. (2003). Occurrence and fate of carbamazepine, clofibric acid, diclofenac, ibuprofen, ketoprofen, and naproxen in surface waters. *Environmental Science and Technology*, **37**(6), 1061-1068.

Vymazal, J. (2007). Removal of nutrients in various types of constructed wetlands. *Science of the Total Environment*, **380**(1-3), 48-65.

Westerhoff, P. and Pinney, M. (2000). Dissolved organic carbon transformations during laboratory-scale groundwater recharge using lagoon-treated wastewater. *Waste Management*, **20**(1), 75-83.

Xue, S., Zhao, Q., Wei, L. and Ren, N. (2009). Behavior and characteristics of dissolved organic matter during column studies of soil aquifer treatment. *Water Research*, **43**(2), 499-507.

CHAPTER 7

EFFECTS OF TEMPERATURE AND REDOX CONDITIONS ON ATTENUATION OF BULK ORGANIC MATTER, NITROGEN, PHOSPHORUS AND PATHOGENS INDICATORS DURING MANAGED AQUIFER RECHARGE[5]

SUMMARY

Laboratory-based soil column and batch experiments simulating soil aquifer treatment were conducted to examine the influence of temperature variation and oxidation- reduction (redox) conditions on removal of bulk organic matter, nutrients and indicator microorganisms using primary effluent. While an average dissolved organic carbon (DOC) removal of 17.7% was achieved in soil columns at 5°C, removal increased by 10% for every increase in temperature by 5°C over the range of 15°C to 25°C. Furthermore, soil column and batch experiments showed higher DOC removal under aerobic (oxic) conditions compared to anoxic conditions. Aerobic soil columns exhibited DOC removal 15% higher than that in the anoxic columns, while aerobic batch showed DOC removal 7.8% higher than the corresponding anoxic batch experiments. Ammonium-nitrogen removal greater than 99% was observed at 20°C and 25°C, while 89.7% was removed at 15°C, but the removal decreased substantially to 8.8% at 5°C. While ammonium-nitrogen was attenuated by 99.9% in aerobic batch reactors at room temperature, anoxic experiments under similar conditions resulted in 12.1% ammonium-nitrogen reduction, with corresponding increase in nitrate-nitrogen and decrease in sulfate concentrations.

[5] Based on Abel et al. (2012). *Water, Air and Soil Pollution* **223**, 5205-5220

[5] Based on Abel et al. (2012). Proceedings of IWA World Water Congress. September 16-22 Busan, South Korea.

7.1 INTRODUCTION

Attenuation of wastewater-derived contaminants is the key objective in any managed aquifer recharge system before it is reused. Removal of suspended solids, organic compounds, nitrogen, phosphorus, bacteria and viruses in a soil infiltration treatment system is achieved through infiltration, percolation, sorption, chemical reaction, biotransformation, die-off and predation (Kanarek and Michail, 1996). Phosphorus removal in soil media is influenced by the physical-chemical properties of the media, and is mainly by sorption and precipitation (Vohla et al., 2007). Organic matter is predominantly removed through sorption and biodegradation in soil aquifer treatment (SAT) system. These removal mechanisms take place simultaneously in the system and part of the organic matter adsorbed onto soil media undergoes biological degradation (Xue et al., 2009). Degradation of oxidizable organic substances is facilitated by microbial metabolism (Greskowiak et al., 2005). Heterotrophic bacteria consume and oxidize organic substances to create new cell material and generate energy for growth and maintenance (van der Aa et al., 2011). Heterotrophic bacteria use dissolved oxygen (DO) to biodegrade natural organic matter (NOM) (van der Kooij et al., 1982). Removal of DO in the unsaturated zone promotes anoxic conditions in the subsequent saturated zone (Fox et al., 2001). Furthermore, seasonal variations affect redox processes in groundwater (Massmann et al., 2006). Bacteria are the most commonly found microbial pathogens in recycled water (Toze, 1999). Nevertheless, reduction of microbial pathogens at tertiary treatment level (i.e. SAT) alleviates potential health risks associated with reuse of poorly treated wastewater (Costán-Longares et al., 2008). *Escherichia coli* (*E. Coli*) and *coliform* bacteria are used as indicators to detect water contamination with faecal pathogens (Kretschmer et al., 2000; Pescod, 1992).

The performance of SAT system is highly influenced by several factors such as wastewater effluent quality, hydrogeological aspects of the site and process conditions applied including; pre- and post-treatment, hydraulic loading rates, and wetting and drying cycles (Amy and Drewes, 2007; Sharma et al., 2008; Sharma et al., 2011).

High ammonium concentration, low nitrate and relatively high phosphorus are major characteristics of primary effluent (Ho et al., 1992). Nonetheless, there is little data available on the mechanism of removal of different contaminants from primary effluent in SAT systems and the interaction between aquifer materials and contaminants. Furthermore, removal of bulk organic matter, nitrogen, phosphorus and pathogenic microorganisms at different temperatures and redox conditions is not well probed. Therefore, the objective of this chapter was to assess the impact of temperature and redox conditions on attenuation of bulk organic matter, nitrogen, phosphorus and pathogens surrogates in laboratory-based soil columns and batch experiments simulating SAT systems.

7.2 MATERIALS AND METHODS

7.2.1 Source water characteristics

Primary effluent (PE) used in this study was collected from Hoek van Holland wastewater treatment plant (WWTP). Characteristics of the PE are presented in section 4.2.1.

7.2.2 Experimental setups

7.2.2.1 Soil column

Four double-walled soil columns (XK50/30; Amersham Pharmacia Biotech, Sweden) detailed in Figure 7.1 were wet-packed by allowing the silica sand grains (0.8 mm - 1.25 mm) to settle in deionized water to ensure homogeneous media packing in the column. The porosity and uniformity coefficient (Cu = d_{60}/d_{10}) of the porous media were 0.4 and 1.3, respectively. Two columns were used to explore the impact of oxidation-reduction (redox) conditions, while the third column was used to assess the influence of temperature variation at 15°C and the fourth column was operated at 5°C. Furthermore, the soil column used to examine aerobic conditions was used to conduct continuous loading experiments at 20 and 25°C. A variable speed peristaltic pump was connected to each soil column using tygon flexible tubes to continuously feed the primary effluent at constant hydraulic loading rate (HLR) of 0.625 m/d from brown glass bottles to the top part of the column after which it percolated down the column under gravity. The columns were operated in down-flow mode and were therefore assumed to be predominantly unsaturated. The empty bed contact time (EBCT) for the soil columns was 11.5 h. Effluent samples were collected from a sampling port situated at the bottom part of the column at least 15 h after the introduction of feed water to the column to ensure enough retention time and consequently sufficient contact between the water and soil media.

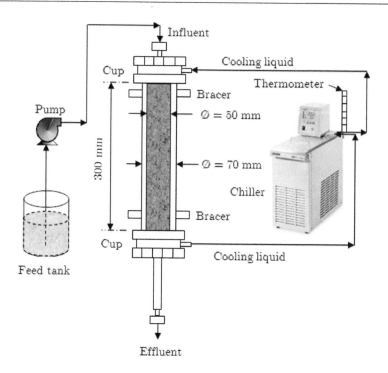

Figure 7.1 Schematic of soil column experimental setup

In order to establish a robust sessile microbial population in the biofilms around the sand media and ensure steady state conditions (biostability) in the columns, the columns were acclimated by continuous loading of the primary effluent and analyzing the DOC of the influent and effluent samples. Operating conditions during acclimation process of the system were maintained aerobic through aeration of feed water at room temperature until DO reached around 8.0 mg O_2/L after which aeration process was discontinued and application of PE to columns started. The columns were assumed to have reached biostability when a difference of less than $\pm 1\%$ DOC removal was obtained between successive samples measurements. This process was repeated for each new experimental operating condition to ensure that the microorganisms in the biofilms around the media had adapted to their new environmental conditions.

To investigate the effect of temperature variation in the simulated SAT system, an ecoline immersion thermostat chiller (E100 w RE106, Lauda Dr. R. Wobser GmbH & CO. KG, Germany) filled with a cooling liquid (Halfords B.V., Netherlands) was connected to the soil column and the outer part of the column was used to circulate the liquid to control the test temperature. A thermometer was plugged (perpendicular to liquid flow) into the outlet tube connecting the chiller with the column. Furthermore, the temperature of the water exiting the column was regularly checked using a 330i handheld conductivity meter (Wissenschaftlische-Technische Werkstatten GmbH & CO. KG, Germany) equipped with an automatic temperature compensation feature and the temperature of the cooling liquid was adjusted accordingly to maintain the required temperature in the system. The system was then run at 15°C, 20°C and 25°C for 2-3 weeks to acclimatized the microorganisms in biofilms around sand particles to the new environment and attain a stable DOC removal. However, low temperature experiments were conducted by placing the entire soil column setup (column, feed water tank and pump) in a cold room under controlled temperature of 5°C during which the outer part of the column was filled with the cooling liquid and the opening ports used to circulate the cooling liquid were sealed.

On the other hand, anoxic conditions were created by inserting a fine stream of nitrogen gas into the feed water tank to strip out DO from the influent water. The tank was filled with influent water (no space above water) to ensure that all oxygen escaped the system. The nitrogen stream was turned off when DO concentration of the feed water achieved <0.2 mg O_2/L after which the feed tank was sealed while maintaining the suction pipe (inside the tank) connected to the pump. DOC concentrations were monitored at influent and effluent points of the soil column operated under anoxic conditions until a difference of <1% was attained between successive DOC measurements. Furthermore, oxidation-reduction potential (ORP) was measured using a redox potential probe connected to calibrated pH meter in millivolt (mV) to crosscheck the anoxic conditions in feed and effluent samples. A minimum of four sample repetitions (n=4) were collected from the biostable soil columns at various operating conditions.

7.2.2.2 Batch reactors

To simulate saturated flow conditions in SAT system at laboratory-scale, 15 glass bottles (0.5 L) in triplicate were filled with 100 g of silica sand (similar to the one used in soil columns above) and fed with 400 mL PE every five days. The reactors were placed on a horizontal reciprocal table shaker and agitated at 100 rpm shaking speed throughout the entire experimental period. Establishment of a sessile microbial population on the sand and steady state operating conditions were achieved in the reactors through determination of removable (biodegradable) DOC as the difference between initial DOC concentrations on day 0 (DOC_0) and final day 5 (DOC_5) over 4 - 6 weeks (data not provided). Aerobic conditions were achieved through aeration of the influent water until DO concentration was 8.0 mg/L. Furthermore, the reactors were considered to have reached steady state operating conditions (biostability) when a robust attached microbial population was established in the biofilms environment on sand grains. DOC removal in soil columns and batch reactors was used to indicate biostable conditions in the reactors with ±1% difference in each three consecutive DOC removal as threshold.

To investigate the influence of anoxic conditions on removal of bulk organic carbon, nutrients (nitrogen and phosphorus) and pathogens indicators (surrogates), six reactors were switched to anoxic operating conditions by stripping out DO using fine stream of nitrogen gas. The nitrogen gas was introduced via a plastic tube (tygon, Saint-Gobain Corporation, France) that penetrates a cap deep into the influent water while the oxygen was allowed to escape the space above influent water through another tube that protrudes from the cap. The reactors were tightly capped with screw-type lids when oxygen concentration reached <0.2 mg/L in the water. Continuous measurement of DO was carried out in the anoxic batch reactors using a HQ30d meter and LDO101 probe (Hach, Colorado, USA) equipped with data log function at 15 minutes time interval for five days to monitor the change in DO over the retention time when DO concentration on day five exceeded 0.2 mg/L. The meters had low and high DO detection limits of 0.01 and 20 mg O_2/L, respectively. The reactors were tightly capped and the probe penetrated down into the water from a hole on the cap which was sealed using paraffin. Monitoring of DOC was carried out at a five day intervals until the reactors were biostable. Samples from the biostable batch reactors were collected (in duplicate) at least three times.

7.2.3 Analytical methods

Analytical methods used to analyze DOC, ultraviolet absorbance at 254 nm (UVA$_{254}$), ammonium-nitrogen (NH$_4$-N), nitrate-nitrogen (NO$_3$-N), phosphate-phosphorus (PO$_4$-P), DO and pathogens indicators were similar to that used in section 3.2.4. Sulfate (SO$_4{}^{2-}$) was measured using an ion chromatography system (ICS-1000, Dionex Corporation, USA). ORP was measured using a redox potential probe connected to a 691 pH Meter (Metrohm, USA).

7.3 RESULTS AND DISCUSSION

7.3.1 Influence of temperature on contaminants removal in soil columns

7.3.1.1 Bulk organic matter

Figure 7.2 shows removal of DOC (n = 4) in soil column experiments conducted at 5°C, 15°C, 20°C and 25°C under aerobic conditions. The percentage and magnitude of DOC removed (expressed as ΔDOC) increased with an increase in temperature. The soil column removed DOC by 17.7±6.4% (ΔDOC = 3.9±1.6 mg/L) at 5°C, 34.3±0.5% (ΔDOC = 9.7±1.0 mg/L) at 15 °C, 45.5±0.3% (ΔDOC = 12.8±1.1 mg/L) at 20°C, and 54.5±0.4% (ΔDOC = 17.1±0.8 mg/L) at 25°C.

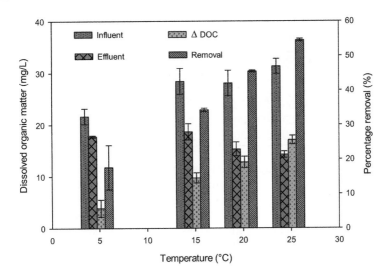

Figure 7.2 Summary of change in DOC concentration with temperature in aerobic soil column experiments fed with primary effluent (EBCT = 11.5 h)

Furthermore, SUVA values proportionally increased with increase in temperature by 9.5 ± 5.7, 39.4 ± 2.3, 46.8 ± 4.9 and $69.6\pm9.1\%$ at 5, 15, 20 and 25°C, respectively. This significant increase ($P < 0.0001$) in DOC removal and increase in SUVA values with an increase in temperature implies that biodegradable substances were preferentially removed due to increase in microbial activity with an increase in temperature. According to Gruenheid et al. (2005) non-degradable DOC increases during cold winter months at recharge sites. Low DOC removal at 5°C suggests that the activity of heterotrophic bacteria that rely on DOC for growth and metabolism was severely affected by low temperature, leading to lower DOC removal and subsequently higher DOC concentration exiting the soil column compared to 15, 20 and 25°C columns.

Fluorescence intensities were identified in three regions in 3D F-EEM spectra. These intensities were differentiated based on the range of excitation and emission wavelengths at which they occurred. Humic-like peak (P1) was observed in wavelength range of ($\lambda_{Ex/Em}$ = 240-250/430-450), fulvic-like peak (P2) covered wavelength range of ($\lambda_{Ex/Em}$ = 320-330/420-430) and protein-like peak (P3) appeared at the wavelength range of ($\lambda_{Ex/Em}$ = 270-280/308-330). As shown in Table 7.1, fluorescence intensities revealed 2% reduction in P1, while P2 remained unchanged at 5°C. Additionally, reduction in P1 was in the range of 8-12% at 20 and 25°C while P2 exhibited reduction of 5.3-8.6% at 20 and 25°C, while P3 was reduced by 26.3%, 26.9% and 50.1% at 5, 20 and 25°C, respectively. However, higher reduction in P1 (35.4%), P2 (27.4%), and P3 (33.0%) were observed at 15°C.

Table 7.1 Change in intensity of characteristic fluorescence peaks at different temperatures in soil column experiments (EBCT = 11.5 h)

Peak	λex/em (nm)	Reduction in intensity of fluorescence peaks (%)			
		5°C	15°C	20°C	25°C
P1 (humic-like)	240-250/430-450	2.0	35.4	7.8	12.0
P2 (fulvic-like)	320-330/420-430	0.0	27.4	8.6	5.3
P3 (protein-like)	270-280/308-330	26.3	33.0	26.9	50.1

Humic substances are broadly defined as large negatively charged molecules that constitute humic and fulvic acid (Laangmark et al., 2004). While these substances resist biodegradation due to their hydrophobicity, they could be removed through adsorption in the subsurface environment (Quanrud et al., 1996). Silica sand used as media in column experiments conducted at 5, 15, 20 and 25°C may had undergone humification during the ripening period as a result of deposition of high molecular (>20,000 Dalton) humic substances leading to exhaustion of adsorption sites within the media. According to Xue et al. (2009) reduction in protein-like substances in soil column experiments simulating SAT is attributed to breakdown of these substances into non-fluorescent structures. High reduction of protein-like fluorescence intensities at 25°C could be ascribed to degradation of fluorescent materials due to increase in microbial activity pertaining to increase in temperature. Schreiber et al. (2007) stipulated that the adsorption process generally decreases at higher temperature. Additionally, high reduction in peak 3 is ascribed to possible increase in microbial activity at 25°C compared to 5°C.

7.3.1.2 Nitrogen

NH_4-N concentration was attenuated fairly significantly ($P < 0.0001$) at high (15°C, 20°C and 25°C) temperatures. However, NH_4-N removal rates decreased considerably at 5°C. An average reduction of 8.8% (28.3±3.7 mg N/L to 25.8 mg N/L) was achieved at 5°C, 89.7% (32.5±5.9 mg N/L to 3.36±1.1 mg N/L) at 15°C, while >99% NH_4-N was reduced at 20 and 25°C. A decrease in NH_4-N concentration resulted in an increase in NO_3-N concentration which marginally increased from 0.3±0.1 to 0.6±0.1 mg N/L at 5°C (Figure 7.3). Nonetheless, NO_3-N concentration increased from less than 2 mg N/L in influent samples to 25.0±2.9, 20.4±1.4 and 23.3±1.1 mg/L at 15, 20 and 25°C respectively.

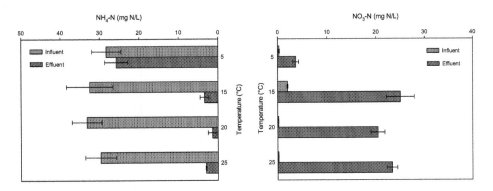

Figure 7.3 Impact of temperature variation on nitrogen removal from primary effluent in soil column experiments under aerobic conditions (EBCT = 11.5 h)

Low NH_4-N removal and corresponding slight increase in NO_3-N at 5°C implied that nitrifying bacteria were affected by low temperature. However, these bacteria remained active and provided limited removal of NH_4-N. This is contrary to the findings of Yamaguchi et al. (1996) who stipulated that growth of nitrifying bacteria ceases at 10°C. Furthermore, NH_4-N removal was 10% less at 15°C than that at 20°C and 25°C. Conversely, NO_3-N concentration in effluent samples from 15°C soil column was somewhat higher than that of 20°C and 25°C columns. These results suggest that another removal mechanism (presumably adsorption) contributed to relatively high removal of NH_4-N at high temperatures due to higher molecular activity that increased collision between positively charged NH_4-N ions and negatively charged silica sand. These results diverge from the findings of Yamaguchi et al. (1996) who observed a decrease in NH_4-N adsorption at 30°C compared to 10°C in a soil column filled with granite mixed with clayey silt.

7.3.1.3 Phosphorus

Mass loading rate (MLR) for PO_4-P at various soil columns operated at different temperatures ranged from 4.9±0.8 to 8.0±0.1 mg/d. Figure 7.4 shows change in PO_4-P concentrations at feed and effluent points of the soil columns at different water temperatures. Phosphorus is poorly removed in sandy soils (Reemtsma et al., 2000). Reduction of 4.9±0.7% (0.4±0.0 mg/d) was observed in PO_4-P concentration at 5°C, 11.4±2.3% (0.8±0.1 mg/d) at 15°C while 23.2±4.7% (1.0±0.2 mg/d) PO_4-P was removed at 20°C and 31.5±1.2% (2.1±0.3 mg/d) removal was achieved at 25°C. These results indicate that higher PO_4-P removal took place at higher ambient temperature.

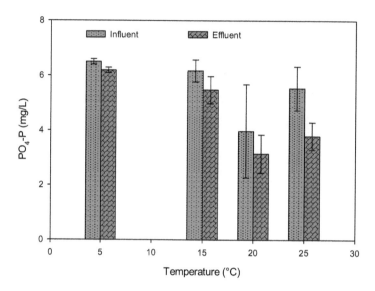

Figure 7.4 Impact of temperature variation on the removal of phosphorus from primary effluent in soil column (EBCT = 11.5 h)

PO_4-P is mainly sorbed or precipitated in filter media (Vohla et al., 2007). Since dissolved organics are removed through adsorption and biodegradation (Idelovitch et al., 2003), removal of PO_4-P might be affected by high organic carbon content in PE through competetion for adsorption sites. However, part of the organic matter adsorbed onto soil particles, is biologically degraded (Idelovitch et al., 2003). PO_4-P adsorption may be negatively affected by organic carbon in PE or through blocking of adsorption sites and/or competition with phosphorus for adsorption sites within the soil particles (Sakadevan and Bavor, 1998). Experiments conducted at 20°C preceded those at 25°C, but the latter exhibited higher phosphorus removal. Increasing PO_4-P removal at high temperature may be due to an increase in degradation of the sorbed materials or increase in degradation of organic matter at 25°C compared to 20°C. Furthermore, increase in temperature might have increased assimilation of PO_4-P adsorption by bacterial biofilms following easy diffusion of PO_4-P into biofilms environment due to decrease in water viscosity at high temperatures. Nonetheless, the role of direct adsorption of PO_4-P to binding sites within silica sand is assumed to be minimal due to negative charges on both the sand and PO_4-P.

7.3.1.4 E. coli and total coliforms removal

Mean concentrations of indicator microorganisms in PE used ranged from 5.9×10^6 to 6.7×10^6 CFU/100 mL for *E. coli* and 19.9×10^6 to 37.3×10^6 CFU/100 mL for *total coliforms*. These indicator organisms were attenuated in the soil column under different operating temperature (Figure 7.5). Both microorganisms were equally removed by 0.36 $\log_{10}$ units (n=6) at 5°C. However, removal of both microorganisms increased at 15°C and *E. coli* was removed by 2.5 $\log_{10}$ units (n=6) while total coliforms showed removal of 2.7 $\log_{10}$. Furthermore, removal of *E.coli* was identical and exhibited no change at 20 and 25°C (3.1 $\log_{10}$ units) (n=6), while *total coliforms* removal increased from 2.9 $\log_{10}$ units (n=6) at 20°C to 3.3 $\log_{10}$ units at 25°C.

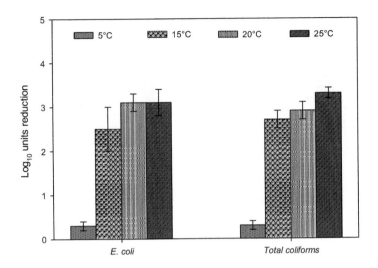

Figure 7.5 Reduction of pathogens indicators ($\log_{10}$ units) at different temperatures in soil column using PE (EBCT = 11.5 h)

Biomass in a low temperature environment excretes extracellular polymers substances (EPS) which enhance clogging of a biofilter (Characklis, 1973; Le Bihan and Lessard, 2000). Presence of EPS in high amounts could adversely affect adsorption of bacteria through blockage of adsorption sites.

7.3.2 Influence of redox on contaminants removal in soil columns

7.3.2.1 Bulk organic matter

DOC removals of 46.4±2.0% (34.3±6.0 to 19.6±4.5 mg/L) and 31.3±0.3% (29.4±3.8 to 19.7±2.4 mg/L) were attained in acclimated soil column experiments under aerobic and anoxic conditions respectively. SUVA values increased from 1.8 to 2.2 L/mg. m under aerobic and from 2.1 to 2.6 L/mg. m under anoxic conditions. These results show that removal under aerobic conditions was 15% higher than that under anoxic conditions. This significant ($P < 0.0001$) difference is attributed to presence of oxygen which is used by the microorganisms to degrade the organic matter in aerobic soil column experiments. On the other hand, DOC removal under anoxic conditions could be ascribed to combination of anoxic degradation, adsorption and presumably aerobic biodegradation in aerobic pockets along the media depth. These results are consistent with the findings of Gruenheid et al. (2005) who observed slower degradation of DOC in anoxic zone compared to aerobic zone during soil passage at a bank filtration site. Increase in SUVA values under both redox conditions indicated that aliphatic substances were removed in both aerobic and anoxic soil columns.

Results of F-EEM (Figure 7.6 and Table 7.2) showed reduction in P1, P2, and P3 under aerobic conditions while the same peaks increased under anoxic conditions.

(a)   (b)

(c)   (d)

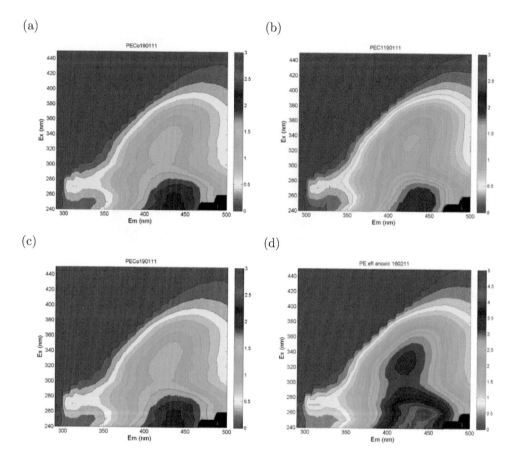

Figure 7.6 F-EEM spectra of soil column studies with PE under different redox conditions: (a) and (c) influent, (b) effluent aerobic and (d) effluent anoxic (EBCT 11.5 h)

Table 7.2 Change in fluorescence peaks intensities of (organic matter) of PE in aerobic and anoxic in soil columns (silica sand, EBCT = 11.5 h)

Redox condition	Intensity of Fluorescence Peaks		
	P1 ($\lambda_{ex/em}$ = 240-250/430-440 nm)	P2 ($\lambda_{ex/em}$ = 320-330/420-430 nm)	P3 ($\lambda_{ex/em}$ = 270-280/310-320 nm)
Aerobic			
Influent	2.28	1.59	0.75
Effluent	2.10	1.49	0.55
Change (%)	7.8 (-)	8.6 (-)	26.9 (-)
Anoxic			
Influent	2.28	1.59	0.75
Effluent	3.45	2.55	0.90
Change (%)	92.4 (+)	111.3 (+)	57.3 (+)

These results are in agreement with the findings of Saadi et al. (2006) monitoring effluent dissolved organic matter (DOM) in the Haifa WWTP effluent using fluorescence, UV and DOC. The authors ascribed decrease in fluorescence intensities to degradation of fluorescing material or dampening of DOM fluorescence by newly formed organic molecules. However, they postulated that increase in fluorescence intensities was due to formation of new fluorescence material affiliated with DOM biodegradation and/or degradation of organic components able to quench fluorescence. Results of a study carried out by Maeng et al. (2008) using soil columns (HRT = 5 days retention time) under anoxic conditions did not reveal increase in fluorescence intensities for P1 and P2. Anoxic conditions in this research work appear to have induced increase in fluorescence intensities for the same peaks in soil column experiments with shorter HRT (11.5 h). This increase in P1 and P2 could be ascribed to humification caused by continuous loading of humic-like fractions at rates higher than the removal in column.

7.3.2.2 Nitrogen

NH_4-N decreased by 99.5±0.2% NH_4-N (33.2±3.8 to 0.2±0.0 mg N/L) in soil column operated under aerobic conditions, while anoxic conditions removed NH_4-N by 71.8±2.0% (32.2±3.8 to 9.3±0.7 mg N/L). Decrease in NH_4-N concentrations yielded substantial increase in NO_3-N concentrations from 0.2±0.0 mg N/L in the influent of the aerobic column to 20.4±1.4 mg N/L in the effluent whereas, NO_3-N concentration increased from 1.9±0.1 to 16.3±2.7 mg N/L in effluent of anoxic soil column samples. Furthermore, an average DO concentration of 0.46 mg/L and ORP of 38.2±1.3 mV were measured in water samples exiting the anoxic column. Figure 7.7 shows change in NH_4-N and NO_3-N concentrations under aerobic and anoxic conditions.

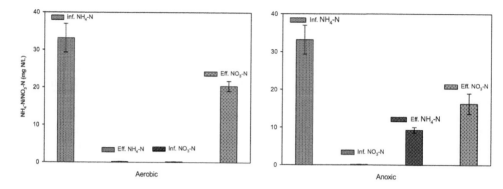

Figure 7.7 Change in nitrogen concentration of PE in soil column operated at room temperature under different redox conditions (EBCT = 11.5 h)

High reduction in NH_4-N concentration and corresponding increase in NO_3-N concentration in influent and effluent samples taken from aerobic soil column indicate that the nitrification process took place in the column. However, decrease in NH_4-N concentration in anoxic samples might be due to adsorption of NH_4-N by soil media or nitrification in aerobic pockets along the column depth resulting from oxygen entrapment in the media. Removal of NH_4^+ under anoxic conditions in presence of organic matter was also observed in previous study conducted by Sabumon (2007) in a 700 mm high and 50 mm diameter column (continuous reactor) seeded with digested cow dung.

7.3.2.3 Phosphorus

MLR of 4.9 ± 0.8 mg/d was applied to aerobic and anoxic soil columns. PO_4-P removal in aerobic soil column was $23.2\pm4.7\%$ (1.0 ± 0.2 mg/d) while anoxic conditions resulted in $22.5\pm2.9\%$ (1.1 ± 0.3 mg/d) PO_4-P removal. This comparable ($P = 0.74$) PO_4-P removal under both aerobic and anoxic conditions in soil columns implies that redox conditions did not affect PO_4-P removal.

7.3.3 Influence of redox operating conditions on removal of contaminants in batch experiments

7.3.3.1 Bulk organic matter

Aerobic biostable batch reactors exhibited DOC removals of $54.5\pm0.3\%$ ($DOC_0 = 35.4\pm2.5$ mg/L and $DOC_5=16.1\pm1.2$ mg/L). However, anoxic batch reactors showed DOC removal of $46.7\pm0.7\%$ ($DOC_0 = 35.4\pm2.5$ mg/L and $DOC_5 = 19.1\pm3.0$ mg/L). SUVA values increased from 1.65 L/mg. m in aerobic water samples to 2.87 L/mg. m whereas SUVA of anoxic water samples increased from 1.65 to 2.68 L/mg. m. Increasing SUVA values during aerobic soil passage were observed by Gruenheid et al. (2005) in a full-scale bank filtration study to monitor DOC. The authors attributed this increase to preferential removal of aliphatic carbon substances. However, increase in SUVA values under anoxic conditions suggests anoxic degradation of aliphatic substances. This is contrary to the findings of the same authors who revealed decrease in SUVA in the anoxic zone. On the other hand, results of F-EEM (Figure 7.8) for aerobic and anoxic experiments showed higher reduction of P1 by 27% in aerobic tests compared to 23% in anoxic batch experiments, while P2 exhibited less reduction of 15% in aerobic batch samples compared to 17.3% in anoxic samples. However, P3 was substantially reduced by 61.5% under aerobic conditions and anoxic batch experiments revealed 38.9% reduction. Correspondence of high P3 reduction to high DOC removal under aerobic conditions implies that biodegradation has an influential role in reduction of protein substances. However, slight differences between reductions of P1 and P2 suggest that adsorption could be the mechanism that dominated the reduction of these peaks.

(a)   (b)

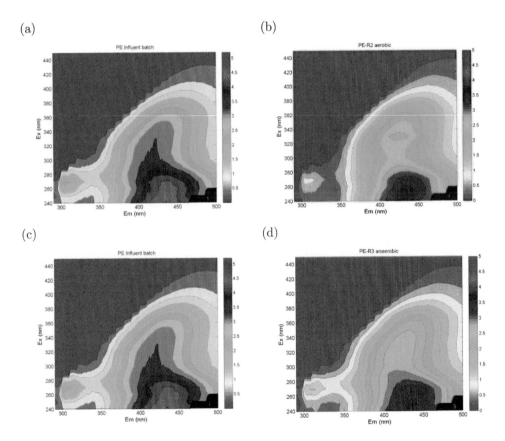

(c)   (d)

Figure 7.8 F-EEM spectra of PE during batch experiments under redox conditions: (a) and (c) influent, (b) effluent aerobic and (d) effluent anoxic (influent: day 0, effluent: day 5)

7.3.3.2 Nitrogen

An average removal of 99.9% NH_4-N (influent = 31.7 ± 4.7 mg N/L) resulted in substantial increase in NO_3-N concentration from 1.9 ± 0.1 mg N/L in influent samples to 36.0 ± 1.8 mg N/L in the effluent under aerobic conditions causing DO reduction from 8 to 2.1 ± 0.6 mg/L. However, NH_4-N decreased by 12.1% (31.7 ± 4.7 to 27.9 ± 1.8 mg N/L) under anoxic conditions causing NO_3-N to increase from 1.9 ± 0.1 to 4.9 ± 1.1 mg N/L in influent and effluent, respectively. Furthermore, ORP values of 199 ± 9.8 mV and 27.4 ± 8.1 mV were measured under aerobic and anoxic conditions. According to Tebbutt (1982), aerobic reactions show ORP values of >+200 mV, while anaerobic reactions occur below +50 mV. A higher ($P < 0.0001$) decrease in NH_4-N concentration in aerobic batch reactors in the presence of organic matter is ascribed to its nitrification to NO_3-N by heterotrophic nitrifying bacteria due to the prevalence of high oxidation conditions manifested by the presence of high DO concentration and ORP in influent water samples. However, lower DO and ORP in effluent samples suggested the prevalence of anoxic conditions. Furthermore, the slight decrease observed in NH_4-N concentration and corresponding increase in NO_3-N concentration under anoxic conditions implies that some nitrification took place. Such anoxic nitrification of ammonium has been reported by Sabumon (2007) who conducted batch studies to assess anaerobic ammonium removal. The author stipulated that anoxic oxidation of ammonium in the presence of organic matter by mixed bacteria cultures was a result of facultative nitrifiers, methanogenesis and SO_4^{2-} reduction. Removal of NH_4-N under anoxic conditions in this study could be due to ANAMMOX process since NO_3-N concentration in the feed water was relatively low (1.9 ± 0.1 mg N/L) to support denitrification process. Measurements of SO_4^{2-} concentrations in influent (PE) water and effluents from aerobic and anoxic batch reactors showed significant increase in SO_4^{2-} concentration by 13.3 mg/Ll (86.5 ± 2.3 to 99.8 ± 3.1 mg/L) in aerobic reactors and decrease in anoxic reactors by 10.5 mg/L (86.5 ± 2.3 to 76.5 ± 3.1 mg/L). Since oxidation in the absence of oxygen is brought about by reduction of inorganic salts such as sulfate (Samorn, 1996), reduction in SO_4^{2-} concentration in anoxic batch reactors indicated utilization of SO_4^{2-} reduction to promote anoxic NH_4-N nitrification. Furthermore, continuous measurement of DO in the batch reactor exhibited steady increase in DO level on day four and day five from 0.2 mg/L to 0.8 mg/L. The used silica sand was relatively free of iron and manganese oxides. An increase in sulfate concentration in aerobic batch reactors may be ascribed to the release of sulfate bound to organic matter due to degradation of the organic matter.

7.3.3.3 Phosphorus

To probe impact of redox conditions on the removal of PO_4-P in batch experiments with hydraulic retention time (HRT) of 5 days, a maximum MLR of 2.67 ± 01.0 mg/d was applied to batch reactors on day 0. Measurements of PO_4-P in the effluent of batch reactors on day 5 showed no significant difference ($P = 0.237$) between removals in aerobic and anoxic batch experiments. PO_4-P removal of $32.2\pm3.4\%$ (0.86 ± 0.09 mg/d) was achieved under aerobic conditions while it was reduced by $26.5\pm6.2\%$ (0.71 ± 0.16 mg/d) under anoxic conditions. Relatively high PO_4-P removal under aerobic conditions could be attributed to increase in the degradation of previously sorbed materials and subsequent mineralization of the PO_4-P attached on these materials.

7.3.3.4 E. coli and total coliforms removal

E. coli and *total coliforms* were removed in laboratory-based batch experiments with 5 days hydraulic retention time under aerobic and anoxic conditions. Since bacteria survival in soil matrix ranges from a few weeks to a few months (Yavuz Corapcioglu and Haridas, 1984), the removal of *E. coli* and *total coliforms* in batch reactors might be ascribed to adsorption of bacteria strains. *E. coli* was reduced by 3.6 $\log_{10}$ units in aerobic batch reactors and by 2.7 $\log_{10}$ units in anoxic reactors. On the other hand, *total coliforms* was attenuated by 2.9 and 2.1 $\log_{10}$ units under aerobic and anoxic conditions, respectively. Organic matter competes with bacteria for adsorption sites resulting in less bacterial adsorption (Stevik, 2004). High bacterial removal (Figure 7.9) under aerobic conditions as compared to anoxic is attributed to high degradation of organic matter under aerobic conditions, predation or die-off due to competition with heterotrophic microorganisms. However, low natural growth rates of *E. coli* and *total coliforms* as compared to their extinction rates in the reactor might have influenced the removal rates in the system.

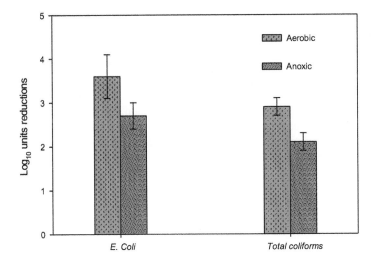

Figure 7.9 Reduction of pathogens indicators ($\log_{10}$ units) under different redox conditions in batch experiments using PE (hydraulic reidence time = 5 days)

The findings of this chapter suggest that removal of bulk organic carbon, nitrogen and indicator pathogens improves significantly at high temperature. Likewise, high contaminants removal was achieved under aerobic operating conditions compared to anoxic conditions, implying that aeration of PE prior to infiltration and application of wetting and drying cycles improve the performance of SAT. Furthermore, succession of aerobic and anoxic conditions during soil passage enhances removal of different contaminants.

7.4 CONCLUSIONS

Assessment of the influence of temperature and redox conditions on removal of bulk organic matter, nutrients and pathogens led to the following conclusions:

Removal of bulk organic matter in aerobic soil columns was significantly higher at high temperature. An increment of 10% was noted for each 5°C increase in temperature in the range 15 to 25°C, while the lowest removal was obtained at 5°C. This was likely due to increasing microbial activity manifested by an increase in SUVA values that exhibited a substantial increase in aliphatic substances removal with increase in temperature. Furthermore, FEEM analysis showed that the protein-like peak was better reduced at high temperature.

Ammonium-nitrogen removal at 15°C was at least 10% less than that at 20°C and 25°C. Nonetheless, experiments carried out at 5°C revealed 8.8% removal suggesting that nitrifying bacteria were sensitive to low temperature levels.

Low removal of *E. coli* and *total coliforms* at 5°C compared to that at 15, 20 and 25°C was likely influenced by competition between these pathogen indicators with poorly removed organic matter and blockage of adsorption sites by EPS secreted by heterotrophic bacteria at low temperature. Inactivation of the pathogen indicators is assumed to have contributed to the removal.

DOC removal was 15% higher in laboratory-scale soil column under aerobic conditions than under anoxic conditions. However, this difference was only 8% in aerobic and anoxic batch experiments. Ammonium-nitrogen concentration was attenuated above 99.9% in aerobic batch reactors while only 12.1% was reduced in anoxic reactors.

The practical implication of these results is that SAT system operated at high temperature in summer will provide high removal rates of DOC, nitrogen, *E. Coli* and *total coliforms* from PE compared to low winter temperature. Inadequate aeration of SAT system due to short drying periods could result in poor reduction of ammonium-nitrogen.

7.5 REFERENCES

Amy, G. L. and Drewes, J. (2007). Soil aquifer treatment (SAT) as a natural and sustainable wastewater reclamation/reuse technology: Fate of wastewater effluent organic matter (EfOM) and trace organic compounds. *Environmental Monitoring and Assessment*, **129**(1), 19-26.

Characklis, W. (1973). Attached microbial growths-I attachment and growth. *Water Research*, **7**, 1113 - 1127.

Costán-Longares, A., Montemayor, M., Payán, A., Méndez, J., Jofre, J., Mujeriego, R. and Lucena, F. (2008). Microbial indicators and pathogens: Removal, relationships and predictive capabilities in water reclamation facilities. *Water Research*, **42**(17), 4439-4448.

Fox, P., Houston, S., Westerhoff, P., Drewes, J., Nellor, M., Yanko, B., Baird, R., Rincon, M., Arnold, R. and Lansey, K. (2001). An investigation of soil aquifer treatment for sustainable water reuse. *Research Project Summary of the National Center for Sustainable Water Supply (NCSWS)*, Tempe, Arizona, USA.

Greskowiak, J., Prommer, H., Massmann, G., Johnston, C., Nützmann, G. and Pekdeger, A. (2005). The impact of variably saturated conditions on hydrogeochemical changes during artificial recharge of groundwater. *Applied Geochemistry*, **20**(7), 1409-1426.

Gruenheid, S., Amy, G. and Jekel, M. (2005). Removal of bulk dissolved organic carbon (DOC) and trace organic compounds by bank filtration and artificial recharge. *Water Research*, **39**(14), 3219-3228.

Ho, G., Gibbs, R., Mathew, K. and Parker, W. (1992). Groundwater recharge of sewage effluent through amended sand. *Water Research*, **26**(3), 285-293.

Idelovitch, E., Icekson-Tal, N., Avraham, O. and Michail, M. (2003). The long-term performance of Soil Aquifer Treatment(SAT) for effluent reuse. *Water Science and Technology: Water Supply*, **3**(4), 239-246.

Kanarek, A. and Michail, M. (1996). Groundwater recharge with municipal effluent: Dan region reclamation project, Israel. *Water Science and Technology*, **34**(11), 227-233.

Kretschmer, N., Ribbe, L. and Gaese, H. (2000). Wastewater reuse for agriculture. *Technology Resource Management and Development-Scientific Contributions for Sustainable Development*, **2**, 37-64.

Laangmark, J., Storey, M., Ashbolt, N. and Stenstroem, T. (2004). Artificial groundwater treatment: biofilm activity and organic carbon removal performance. *Water Research*, **38**(3), 740-748.

Le Bihan, Y. and Lessard, P. (2000). Monitoring biofilter clogging: biochemical characteristics of the biomass. *Water Research*, **34**(17), 4284-4294.

Maeng, S., Sharma, S., Magic-Knezev, A. and Amy, G. (2008). Fate of effluent organic matter (EfOM) and natural organic matter (NOM) through riverbank filtration. *Water Science and Technology*, **57**(12), 1999.

Massmann, G., Greskowiak, J., Dünnbier, U., Zuehlke, S., Knappe, A. and Pekdeger, A. (2006). The impact of variable temperatures on the redox conditions and the behaviour of pharmaceutical residues during artificial recharge. *Journal of Hydrology*, **328**(1-2), 141-156.

Pescod, M. (1992). Wastewater Treatment and Use in Agriculture. Food and Agriculture Organization Irrigation and Drainage Paper 47. Rome

Quanrud, D., Arnold, R., Wilson, L. and Conklin, M. (1996). Effect of soil type on water quality improvement during soil aquifer treatment. *Water Science and Technology*, **33**(10), 419-432.

Reemtsma, T., Gnir , R. and Jekel, M. (2000). Infiltration of combined sewer overflow and tertiary municipal wastewater: an integrated laboratory and field study on nutrients and dissolved organics. *Water Research,* **34**(4), 1179-1186.

Saadi, I., Borisover, M., Armon, R. and Laor, Y. (2006). Monitoring of effluent DOM biodegradation using fluorescence, UV and DOC measurements. *Chemosphere,* **63**(3), 530-539.

Sabumon, P. C. (2007). Anaerobic ammonia removal in presence of organic matter: A novel route. *Journal of Hazardous Materials,* **149**(1), 49-59.

Sakadevan, K. and Bavor, H. (1998). Phosphate adsorption characteristics of soils, slags and zeolite to be used as substrates in constructed wetland systems. *Water Research,* **32**(2), 393-399.

Samorn, M. (1996). Wastewater characteristics. *Resources, Conservation and Recycling,* **16**(1-4), 145-159.

Schreiber, B., Schmalz, V., Brinkmann, T. and Worch, E. (2007). The effect of water temperature on the adsorption equilibrium of dissolved organic matter and atrazine on granular activated carbon. *Environmental Science and Technology,* **41**(18), 6448-6453.

Sharma, S. K., Harun, C. M. and Amy, G. L. (2008). Framework for assessment of performance of soil aquifer treatment systems. *Water Science and Technology,* **57**(5), 941-946.

Sharma, S. K., Hussen, M. and Amy, G. L. (2011). Soil aquifer treatment using advanced primary effluent. *Water Science and Technology,* **64**(3), 640-646.

Stevik, K. (2004). Retention and removal of pathogenic bacteria in wastewater percolating through porous media: a review. Water Research, 38(6), 1355-1367.

Tebbutt, T. (1982). Principles of Water Quality Control. 3rd ed. Pergamon Press, Birmingham, UK.

Toze, S. (1999). PCR and the detection of microbial pathogens in water and wastewater. *Water Research,* **33**(17), 3545-3556.

van der Aa, L., Rietveld, L. and Van Dijk, J. (2011). Effects of ozonation and temperature on biodegradation of natural organic matter in biological granular activated carbon filters. *Water Engineering Science,* **4**, 25-35.

van der Kooij, D., Visser, A. and Hijnen, W. (1982). Determining the concentration of easily assimilable organic carbon in drinking water. *Journal American Water Works Association,* **74**(10), 540-545.

Vohla, C., Alas, R., Nurk, K., Baatz, S. and Mander, Ü. (2007). Dynamics of phosphorus, nitrogen and carbon removal in a horizontal subsurface flow constructed wetland. *Science of the Total Environment,* **380**(1-3), 66-74.

Xue, S., Zhao, Q., Wei, L. and Ren, N. (2009). Behavior and characteristics of dissolved organic matter during column studies of soil aquifer treatment. *Water Research,* **43**(2), 499-507.

Yamaguchi, T., Moldrup, P., Rolston, D. E., Ito, S. and Teranishi, S. (1996). Nitrification in porous media during rapid, unsaturated water flow. *Water Research,* **30**(3), 531-540.

Yavuz Corapcioglu, M. and Haridas, A. (1984). Transport and fate of microorganisms in porous media: A theoretical investigation. *Journal of Hydrology,* **72**(1-2), 149-169.

CHAPTER 8

FRAMEWORK FOR SITE SELECTION, DESIGN, OPERATION AND MAINTENANCE OF SOIL AQUIFER TREATMENT (SAT) SYSTEM

SUMMARY

Soil aquifer treatment (SAT) as a means for wastewater reclamation and reuse is increasingly becoming popular in developed and developing countries faced with severe water shortages. However, absence of a clear guidelines and easy tools is hindering application of SAT in some parts of the world. Several tools were developed in this study, which can be used during different stages of SAT scheme. A simple tool was developed for the initial planning stages of SAT, which was based on five aspects including institutional, legal and regulatory, socio-political, economical and technical. This tool serves as a preliminary indicator for feasibility of a SAT scheme that could be used by planners and decision makers to informatively decide if on SAT technology feasibility. The following step was development of a site selection tool, which was based on three site specific factors (physical, hydrogeological, land use and economical). This tool was also meant for the planning stage to choose the most suitable site from a number of sites. Coupled with this tool, is a guideline that summarizes site investigation works and laboratory tests involved in development of new SAT scheme. A design guideline was developed for the design stage and it details the important considerations, parameters and steps involved in designing a SAT systems including pre- and post-treatment systems. The operation and maintenance and monitoring guidelines were developed for post-design stage and they present the requirements of a SAT scheme during operation. Furthermore, an Excel-based model was also developed for the design stage to predict contaminants organic matter, nitrogen, phosphorus, bacteria and viruses' removal based on travel distance through SAT in combination with few pre-treatment technologies.

8.1 INTRODUCTION

The concept used in lake and river bank filtration was first used in wastewater effluent reclamation and reuse when the first soil aquifer treatment (SAT) project was developed in Los Angeles County, USA in 1962 using spreading basins. Today, SAT has become a common water reclamation practice worldwide. Generally, successful reclamation and reuse practices require careful planning steps, economic calculations, technical feasibility and detailed social considerations and assessments (Crook and Surampalli, 1996; Huertas et al., 2008). With appropriate planning, water quality control and assessment, groundwater recharge for different reuse purposes can be safely undertaken (Asano and Cotruvo, 2004). According to Drewes (2009), wastewater quality, spreading basin characteristics, degree of blending with native groundwater and operational conditions affect the final recovered water quality. These factors are directly influenced by engineering design of the system. Other factors such as the sub-surface soil characteristics are less determined and dependent upon the individual site and local hydrogeological conditions. This implies that with proper planning, site selection, design, operation and maintenance and monitoring procedures, the successful reclamation of wastewater and its ultimate reuse in various applications can safely be achieved through SAT systems. Furthermore, its successful implementation could be undertaken at identified locations in different regions across the globe.

Nonetheless, absence of a clear guideline that outlines important factors that should be taken into account when considering implementation of SAT technology at potential sites especially in developing countries where it is needed the most, limits its application. Furthermore, the absence of a framework that can guide an engineer through the site selection, design, operation and maintenance and long-term monitoring of SAT systems hinders its propagation.

This chapter aims to address these short-comings on the successful application of SAT so that this technology can be promoted and replicated. The study identifies clearly the factors that influence development of new SAT scheme at different stages and seeks to develop different tools to assist in design, operation and maintenance of the site. Furthermore, it provides a spreadsheet model that predicts the fate of some contaminants during operation of the potential SAT site.

8.2 RESEARCH METHODOLOGY

8.2.1 Desk study

Extensive literature review was conducted whereby journals, articles, books/book chapters, MSc and PhD theses and reports directly or indirectly related to the research subject were reviewed. Relevant data were extracted, compiled and analyzed in order to help build up tools and water quality prediction model, which constitute the work under study. The methodology used to develop SAT scheme is presented in Figure 8.1.

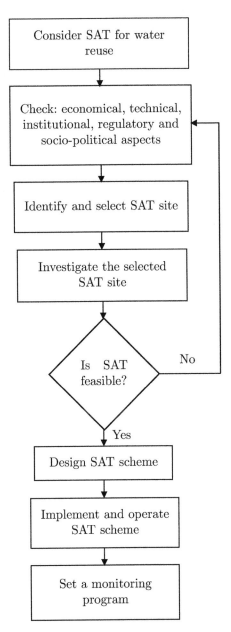

Figure 8.1 Flowchart diagram for development of feasibility study, design and operation of SAT scheme

8.2.2 Development of SAT pre-screening tool

This study was conducted using a quantitative method by distributing a questionnaire survey to generate information for pre-screening of SAT sites that could be used by planners and engineers to develop a new SAT scheme.

Development of the pre-screening tool involved identification of the important aspects that should be addressed during planning stages of a reuse scheme such as SAT and that have led to the success of the schemes under operation. Based on the literatures consulted, five aspects were considered essential for development of the tool using a multi-criteria rating approach. These aspects include institutional, economical, socio-political, technical and legal.

A questionnaire survey was conducted for the pre-screening tool to check the relevance of each criterion. The aim of the questionnaire survey was to establish a methodology that helps planners and engineers rate the factors that influence the successful implementation of reuse project at its early planning stages. Prior to survey distribution, a thorough review of existing studies was carried out to identify some key factors that require scrutiny and an inventory was then developed. In total, 23 factors were initially identified and categorized under institutional, legal, socio-political, economical and technical aspects (see section 2.2 of this thesis). The factors that were shortlisted in the survey were either related or overlapping each other. A detailed description of each factor was provided to enable respondents understand it before sharing their views. The questionnaire survey was distributed to a predefined group of 23 (including 12 MSc students in Sanitary Engineering core and 11 PhD students in department of Environmental Engineering and Water Technology) at UNESCO-IHE Institute for Water Education. The selected group attended lectures on water reuse through managed aquifer recharge as part of their MSc program. Out of this group, 21 students responded.

8.2.3 Tool for site identification, selection and investigation

Development of a tool for site identification and selection involved identification of three key factors (physical, hydrogeological and land use and economical) from the literature review. This is followed by establishment of field investigation criteria that involves identification of type and number of tests to be conducted. Furthermore, a comprehensive laboratory analysis of the sample collected during field investigation is part of this process to ensure absence of rock outcrops and confining materials.

Development of the site investigation tool involved tabulation of all tests and analysis that should be conducted during the investigation work and their extents and details as found in several literatures.

8.2.4   Tool for design of SAT systems

In developing a tool for design of SAT systems, seven parameters were identified from literature review. These parameters include pre-treatment, HLR, wetting/drying cycles, infiltration area, number of basins and geometry, groundwater mound and travel time.

8.2.5   Tool for SAT operation, maintenance and monitoring

The development of this tool was based on field operation data gathered from different SAT sites under operation. Factors such as type of wastewater effluent and length of wetting/drying cycles of recharge basins were considered. Furthermore, monitoring requirements of a SAT scheme include location and frequency of quality parameters for both wastewater effluent and groundwater.

8.2.6   Modeling contaminants removal during SAT

In developing the tool under this section, several pre-treatment technologies applicable to SAT were considered and data from field and laboratory experiments around the world were used. The contaminants considered in developing this model include carbon (as DOC), nitrogen (as NH_4-N and NO_3-N), phosphate as phosphorus (PO_4-P), bacteria and viruses. Pre-treatment technologies considered include settling ponds, coagulation and roughing filters rapid sand filters, disinfection, micro-filtration and ultra-filtration. The removal efficiencies achieved through these pre-treatment technologies for the above specified contaminants were compiled (Table 8.5). Likewise, the removal efficiencies for the specified contaminants with respect to travel distance during SAT were also compiled in the same manner (Figure 8.4 to Figure 8.6). The data for SAT removal efficiencies based on the travel distance and time were collected from laboratory and field schemes. Using these data, formula were inserted in excel sheet which combines the removal efficiencies of the combination chosen (i.e. between the pre-treatment and SAT). SAT removal efficiencies were grouped based on a range of travel distance (0-5, 5-10, 10-20 and 20-100 m).

8.3   RESULTS AND DISCUSSION

8.3.1   SAT pre-screening tool

A questionnaire-based survey was conducted to analyze different aspects relevant for implementation of SAT system. Respondents were asked to rank pre-defined factors using a four-point scale as either "extremely important = 3", "important = 2", unimportant, but needs to be in place or addressed = 1" and "not relevant = 0". The numbers associated with ranks were assigned to facilitate further analysis of responses. Frequency index (FI) for each factor was computed to rank factors based on their importance using the following formula (Assaf and Al-Hejji, 2006; Le-Hoai et al., 2008):

$$FI(\%) = \frac{\sum a \times n \times 100}{H \times N}$$

(8.1)

Where a is the weight given to each factor by the respondents (ranges from 0 for not relevant to 3 for extremely important); n is frequency of each response; H is the highest ranking available (i.e. 3 in this case) and N is the total number of responses. Table 8.1 presents the FI of the factors that were considered important by the respondents to be addressed during SAT pre-screening phase of any potential SAT scheme.

Table 8.1 Ranking of factors considered important for pre-screening of SAT sites

Category	No.	Factor	FI (%)	Rank
Institutional	I1	Presence of institutional framework for wastewater reuse management, with clear identification of the stakeholders and their roles and responsibilities defined.	82.5	4
	I2	Availability of legal structure reflecting stakeholders' responsibilities.	74.6	12
	I3	Establishment of an executive body such as an inter-agency technical standing committee (reuse committee, responsible for sector development, planning and management)	66.7	21
	I4	Effective involvement of stakeholders and achievement of public acceptance.	87.3	1
	I5	Making sure that institutional agencies have been assessed in terms of staffing and skills requirements.	76.2	10
	I6	Availability of a capacity building program to strengthen weak or inadequate legal and regulatory structures	73.0	16
Legal and regulational	L1	Ensuring the existence of legal regulatory standards and make sure that the regulations (by law) permit such reuse.	77.8	8
	L2	Having a guideline in place for the specific type of reuse in the region	79.4	7

Cont.

Category	No.	Factor	FI (%)	Rank
Socio-political	S1	Involvement of the public and key stakeholders (beneficiaries) during execution stages of the project	82.5	4
	S2	Agreeing on long term organizational commitment to inform and educate the public/community about efficient water use and reuse	81.0	6
	S3	Long term public involvement in the decision making during operation and extension of the project	74.6	12
	S4	Setting up communication channels between the public and the authorities/organization	73.0	16
Economical	E1	Cost of land	85.7	2
	E2	Location of WWTP relative to SAT site (km)	84.1	3
	E3	Pre-treatment requirement (treatment level at WWTP) prior to reclamation	68.3	20
	E4	Location of reclaimed water demand area (point of use) relative to SAT site (km)	71.4	19
	E5	Post-treatment requirement depending on intended use after abstraction	61.9	22
	E6	Duration and cost required to review and approve project's documents	61.9	23
	E7	Availability of reuse market	76.2	10
Technical	T1	Soil Type at the proposed site	74.6	12
	T2	Extraction possibility (recharge location and abstraction point within the same region)	73.0	16
	T3	Depth of vadose zone (m)	77.8	8
	T4	Depth of groundwater table (m)	74.6	12

The ranking of the factors that should be considered in any new SAT system in Table 8.1 indicates that establishment of Effective involvement of stakeholders and achievement of public acceptance (87.3), land cost (85.7), location of WWTP relative to SAT site in km (84.1), involvement of the public and key stakeholders during project execution (82.5), presence of institutional framework for wastewater reuse (82.5), ensure project's documents review and approval within a reasonable duration at cost (81.0), having a guideline in place for the specific type of reuse (79.4), depth of vadose zone (77.8), ensure the existence of legal regulatory standards that permits such reuse type (77.8) and availability of reuse market (76.2) were considered the top 10 significant factors. It was noticed that some factors were equally rated with the same FI. In such case, the factor with higher number of "extremely important" was ranked higher. Table 8.2 presents 10 factors required for establishment of new SAT site as ranked by respondents with the first factor being the most important.

Table 8.2 Ranking of factors to be considered during preliminary stages of SAT site development

Category	Factor	FI (%)	Rank
Institutional	Effective involvement of stakeholders and achievement of public acceptance.	87.3	1
Economical	Cost of land	85.7	2
Economical	Location of WWTP relative to SAT site (km)	84.1	3
Socio-political	Involvement of the public and key stakeholders (beneficiaries) during execution stages of the project	82.5	4
Institutional	Presence of institutional framework for wastewater reuse management, with clear identification of the stakeholders and their roles and responsibilities	82.5	5
Socio-political	Agreeing on long term organizational commitment to inform and educate the public/community about efficient water use and reuse	81.0	6
Legal	Having a guideline in place for the specific type of reuse in the region	79.4	7
Technical	Depth of vadose zone (m)	77.8	8
Legal	Ensuring the existence of legal regulatory standards and make sure that the regulations (by law) permit such reuse type.	77.8	9
Economical	Availability of reuse market	76.2	10

The above results suggest that institutional aspects received high importance and public acceptance of MAR project plays a pivotal role in establishing a new reuse project. This is consistent with Hartley (2006) who demonstrated that failure to gain public acceptance had led to abolishment of a potable water supply project in San Diego, California even after gaining approval from panels of technical experts. Economical aspects were of great importance where availability of land at reasonable cost and proximity of SAT site to WWTP rated the second and the third, respectively. Both factors influence investment and operation costs of reuse project as basin size range from less than 0.4 to more than 8 ha (Crites et al., 2000; Metcalf et al., 2007), whereas distance of SAT site relative to WWTP necessitates longer connection pipes and increases pumping costs leading to low suitability of the potential site. Socio-political factors of stakeholders' involvement during execution of the project and long term commitment to educate the public on efficient water use and reuse were rated in the fourth and sixth place, respectively. Legal aspects of ensuring the availability of regional guidelines for this particular water reuse and confirming that local regulations permits such reuse were ranked in the seventh and ninth place. Technical factor of depth of the vadose zone and economical factor of reuse market availability ranked in the eighth and the tenth.

It could be concluded that economical factors were predominantly the most critical factors while technical aspects were the least influential ones. The results suggest that while economical aspects are bottleneck for a successful implementation of SAT scheme, technical factors could be engineered to meet reclaimed water quality requirements. Nevertheless, the above findings are confined to relatively small group of water professionals from one institution and hence the results obtained should used as guide. Furthermore, the above mentioned factors should be checked carefully at each potential SAT site to avoid data extrapolation based on the findings obtained at any existing site.

8.3.2   Site identification, selection and investigation

After ensuring that institutional, technical, economical and socio-political requirements are fulfilled, the next step would be to identify potential SAT sites. SAT using infiltration basins is a land treatment technology that can be adopted where a suitable site is available. Site selection is regarded here as a planning stage process, because absence of a suitable site nullifies the application of the technology. An improper or insufficient site evaluation has in many instances been the cause of failure of some SAT systems. This makes site selection a very important parameter that dictates the success of a SAT project (Crites et al., 2006; Reed et al., 1985). Site identification and investigation involves gathering available data on potential sites to compare and evaluate through desk study. To identify the potential land treatment sites, it is necessary to obtain data on land use, soil types, topography, geology, groundwater, surface water hydrology, applicable water rights issues (Crites et al., 2000; Crites et al., 2006; USEPA, 2006), precipitation and evaporation data. This is followed by site data verification through further field investigation by conducting trial pits, boreholes, infiltration tests and groundwater wells. Important aspects considered during site selection and investigation are presented in Table 8.3

Table 8.3 Site aspects and tests conducted during site identification, selection and investigation

Activity	Factors considered/tests to be conducted
Site identification and selection	Physical
	Hydrogeological
	Land cost and land use
Site investigation	Trial pits
	Boreholes
	Infiltration test
	Groundwater wells

8.3.2.1 Tool for SAT site identification and selection

Based on the above findings, the summary of the site identification and selection factors and their relevant proposed criteria are listed below:

   a- Physical factors
- The land area available in a selected location must be greater than or equal to the estimated infiltration area required for SAT basins and administration.

- A site with land grade 0-5 % is the most preferable location for SAT and a grade of 5-15 % is considered moderately suitable. However, sites with grades greater than 15 % are not suitable for SAT.
- Sites that are unsusceptible to flooding are the most suitable for SAT, while those that are susceptible to flooding are not suitable, especially if expensive interventions, such as construction of dikes are required.

b- Hydrogeological factors

- A suitable site for SAT should have a minimum vadose zone thickness of 5 m and sites with vadose zone thickness less than 5 m are considered not suitable.
- Sandy soil is the most preferable soil type followed by loamy soil. The site is considered unsuitable if its soil type is clayey.
- A homogeneous soil profile within the vadose zone is the most suitable for SAT. Furthermore, if the soil profile is heterogeneous (i.e. contains clay or silt layers above the permeable layer and the thickness of the clay or silt layer does not exceed 2 m), then the site could also be considered suitable. Nevertheless, soil profile that constitutes clay fractions greater than 10% (by mass) is regarded unsuitable for SAT.
- Soils with permeability ranging from 0.36-1.2 m/d are considered suitable for SAT. Unsuitable sites are those with permeability less than 0.36 m/d.
- A confined aquifer is not suitable for SAT.

c- Land use and proximity of SAT scheme to WWTP

- Agricultural or open spaced lands are the most suitable for SAT sites. Besides, low density residential or urban areas can still be considered as suitable. Nevertheless, high density residential and industrial areas are not suitable.
- Potential SAT site which is 5 km from the wastewater effluent source is the most preferable. However a site in the distance range of 5-20 km is considered less suitable, while a distance of more than 20 km is regarded unsuitable.
- When the elevation difference between SAT site and wastewater effluent source is less than 50 m, the site is regarded as suitable. However, an elevation difference greater than 50 m renders the site unsuitable for SAT scheme.

The result of the site identification and selection process might be any of the following three outcomes:

- No site is suitable for application of SAT via infiltration basins due to lack of sufficient land area and absence of a suitable soil within an excavatable depth. This does not mean that SAT technology cannot be applied as other SAT options such as vadose zone wells or direct injection wells can be explored.
- Two or more sites have similar characteristics. In this case, the sites will require a detailed review and comparison.
- One site is suitable. In this case, a site visit will be required for preliminary investigation works.

8.3.2.2 Tool for SAT site investigation

For both of the last two outcomes, a visit to physically assess and explore the site would be necessary. This is important because it helps to verify the existing data and also identify probable or possible site limitations (Crites et al., 2000). For example, the presence of rock outcrops would mean varying soil thickness whereas wet areas or superficial salt crusts would be signs of drainage problems (USEPA, 2006). In both cases, this will mean that more detailed field investigation would be required.

The most challenging and difficult part of site investigation work is never the type of test required or the specific procedures to follow, but rather the decision on the number of tests required and their suitable placement (Crites et al., 2000).

- Trial pits: for inspection of soil profile, texture, structure and to detect presence of any restricting (low permeability) layers within the shallow part of the vadose zone.
- Boreholes: for exploring soil sections beyond the depth of the trial pit, collecting soil sampling for further laboratory testing's, and determining depth of groundwater level and impermeable layers.
- Infiltration tests: for measuring the expected minimum infiltration rate
- Groundwater wells: for determining thickness of aquifer, horizontal permeability and groundwater quality testing.

Minimum requirements for field tests and sampling for SAT site investigation work are presented in Table 8.4.

Table 8.4 Minimum test and sampling requirements for SAT scheme investigation

Test	Minimum requirement	Sampling/tests required
Test pits	3 tests per site with depth ≤4 m	Soil sampling for laboratory testing, groundwater level monitoring by installing a piezometer if groundwater is struck within the excavated depth.
Boreholes	3 boreholes per site up to 10-15 m	Soil and groundwater samples for laboratory analysis and horizontal hydraulic conductivity.
Infiltration test	A minimum of one basin for every major soil type	Pilot basin test with a minimum area of 7-9 m² to be operated for weeks using the same wetting and drying cycles planned for the full scale basins and same wastewater effluent intended for the SAT.
Groundwater wells	3 wells	Pump test to determine the permeability of the aquifer and water quality analysis.
Groundwater quality	–	Tests include pH, TDS, EC, DO, major ions and trace compounds.
Soil samples	–	Particle size distribution, pH, electrical conductivity, Chloride and leachable metals such as iron (Fe), manganese (Mn), arsenic (As) and cadmium (Cd).

8.3.3 SAT system design

When undertaking the design of an infiltration basin for SAT system, several parameters have to be evaluated and a step wise calculation approach of each parameter should be adopted. These parameters include wastewater effluent pre-treatment, hydraulic loading rate, wetting and drying cycle, infiltration area, basin layout, groundwater mound and travel time.

8.3.3.1 Design procedures for SAT systems

The process design for a SAT system excluding post-treatment as mentioned under section 2.4 consists of six steps, followed by additional three steps when nitrogen removal is a process design consideration. Once a suitable site has been selected during the selection stage, a detailed site investigation and basin infiltration test have been conducted; data gathered would be used for the design stage. The data collected should be sufficient to give a clear understanding of the sub-surface conditions and the groundwater system. Based on this and the above listed considerations, the process design for a SAT system should be undertaken in the following sequence:

1. Determination of the type of pre-treatment
The choice of the pre-treatment is accomplished through characterization of wastewater effluent quality, check of available reuse guidelines and comparison of a number of pre-treatment options in combination with SAT, followed by choosing the most optimum pre-treatment process.

2. Determine the wastewater effluent flow rate
This entails calculation of the total volume of effluent based on population and water consumption, WWTP average daily flow data and forecasted reclaimed water demands. Based on this, a percentage of wastewater effluent is allocated for SAT.

3. Determine wetting/drying cycle
The length of wetting and drying periods is determined based on the pilot test conducted at the site, land availability and operation objectives (maximum removal of nitrogen and biological oxygen demand and recovery of infiltration rates).

4. Determine area requirements
Calculate the average infiltration rate based on field infiltration test, wetting and drying periods and local climatic conditions (i.e. precipitation and evaporation). Furthermore, calculate the nitrogen and biological oxygen demand (BOD) loading rates and check with the permissible loading. Consider additional area for infrastructures and possibility of future expansion.

5. Determine groundwater mound
Calculate the groundwater mounding and check the maximum capillary rise above the highest point on the mound. This should not come within 0.6 m of the basin bottom.

6. Determine number and sizing of basins

This will be based on topography, wetting/drying cycle and groundwater mound. Equation 2.3 can be used to calculate number of basins.

7. Determine location of recovery wells

This will be based on either regulation requirements, residence time to achieve desired water quality improvements or proportion of recharge water that needs to be abstracted as mentioned under section 2.4.7.

8. Determine location of monitoring wells

This is based on the location of recovery wells, other nearby private or public wells, or any nearby potable water supply aquifer as mentioned under section 2.4.7.

9. Determine post-treatment requirements

This will be based on pre-treatment and the predicted water quality of recovered water compared to the reuse water quality requirement as mentioned under section 2.4.9. If the quality of the recovered water does not meet this quality requirements then a suitable post-treatments has to be chosen.

8.3.4 Operation, maintenance and monitoring of SAT systems

Operation and maintenance of infiltration basins are conducted to maintain acceptable levels of infiltration rates and ensure that the treatment process within the vadose zone is not impeded. Restoration of infiltration rates is controlled through application of wetting/drying cycles and periodic cleaning of the basin. Other operational measures include monitoring of effluent height within the basins and depth of groundwater beneath the basin infiltrative surface. Another important parameter worth mentioning is the high turbidity increase that may occur during the first fill of the basins when flow is intermittent. An operational strategy would be to maintain stable flows in the pipeline in order to avoid such turbidity peaks at the start of a wetting period (Miotli ski et al., 2010). Another strategy would be to have a separate pond that receives the first fill effluent and resume flow back to the main basin once flow has been stabilized. SAT site may also consist of a number of pipelines (conveyance and distribution within the site and demand points), mechanical or electro-mechanical fittings and possibly pumps, which need to be maintained properly.

8.3.4.1 Wetting and drying

Wetting and drying are important operation parameters in the O&M of infiltration basins. The length of wetting and drying cycles is dependent on soil characteristics, development of a clogging layer, distance to the groundwater table (USEPA, 2012), type of wastewater effluent applied and the ultimate treatment objective (i.e. nitrification and/or denitrification). Other external factors that affect the operating conditions include temperature, precipitation and solar incidence (Fox et al., 2001a). The length of wetting and drying periods would have already been estimated during the design process. However, each basin on the project may tend to have different behavior in terms of infiltration rates and build-up of clogging layer. Due to effects of seasonal change, it will be necessary to fine-tune the wetting and drying periods during operation. Fine tuning helps to optimize basin operation with respect to increasing the HLR and maximizing nitrogen removal. This operational duty depends on the operators who will gain more understanding of their basins over time. The following are important considerations and precautions for basins operation:

- To restore infiltration rates, sufficient drying periods are necessary so that clogging layer can crack, curl up, desiccate and decompose before the next wetting cycle.
- To maximize nitrification, sufficient drying period is necessary so that air can penetrate deep into the vadose zone.
- Increasing a wetting period should increase the depth at which ammonium is adsorbed while increasing a drying period should increase the depth at which the adsorbed ammonium is nitrified.
- Long wetting periods may lead to large decline in infiltration rates, especially towards the end of the period and this will likely cause long drying periods since basin will take longer time to drain.
- Short wetting cycles disrupt insect life cycle and cut down on the growth of algae. This can be helpful in minimizing the reduction of hydraulic conductivity of the clogging layer.
- Extension of drying and shortening of wetting periods are required in winter periods to cater for the low rate of treatment.

However, it must be noted that deciding on the optimum wetting/drying cycles which will give the highest infiltration rate or the maximum nitrogen removal is quite a complex and difficult operation. This is because it is not only affected by type of effluent and season (temperature), but rather because more external factors affect the length of each cycle. These factors include precipitation and evaporation, pre-treatment of the effluent prior to recharge, soil type, site heterogeneity and the presence of restricting layers within the vadose zone.

8.3.4.2 Maintenance

Once the infiltration rates are no longer recoverable by the wetting and drying cycle, then a maintenance work is required. This would include cleaning the infiltration basins by scraping the top layer of the soil followed by disking. To know when maintenance will be necessary, the drying time required after each wetting cycle should be recorded at the beginning of the basin operation and every other subsequent cleaning operation. Once the length of drying is noticed to exceed the normal trend, basin maintenance will be employed.

8.3.4.3 Wastewater effluent

Monitoring the wastewater effluent quality that is used for recharge is conducted to ensure that the effluent quality meets regulation requirements prior to land application. Knowing the characteristics of the effluent also helps to calculate the loadings on the infiltration basins and when compared with the treated water the performance of the SAT system can be evaluated. The effluent parameters to be measured are usually dictated by the regulations based on the reuse application of reclaimed water. These may include any of the physical, chemical and bacteriological parameters. Depending on the parameter, the monitoring frequency recommended can vary from once a day to twice a year.

8.3.4.4 Monitoring of wastewater effluent depth in infiltration basins

Due to rapid decrease in infiltration rate towards the end of the wetting period, especially when long wetting periods are used, wastewater effluent level in the basin will need to be monitored at least twice per day. This is based on the fact that infiltration rates as a result of clogging usually decrease by 50% (Metcalf et al., 2007). This leads to increase in water depth in the infiltration basin.

8.3.4.5 Monitoring of groundwater mound

Depth of groundwater below the infiltration area should be monitored to ensure that the maximum capillary rise above the highest point on the mound should not come within 0.6 m of the infiltration basin bottom.

8.3.4.6 Monitoring of reclaimed water quality

Water quality monitoring is an essential activity in reclaimed water schemes to ensure that public health and the environment are protected (USEPA, 2012). Regulations or guidelines for water reuse usually dictate the water quality monitoring requirements of a groundwater recharge scheme. It is understood that wastewater effluent quality monitoring ensures that reuse regulations are not breached.

Monitoring of groundwater quality serves two purposes: The first is to study or detect any recent adverse impact caused by scheme to the local groundwater while the second is to assess and provide long-term verification of the integrity of the system. The advantage to this as stated by the author, is that public perception is improved along with gain of consumer confidence, especially when the integrity of the system is verified.

8.3.4.7 Monitoring of groundwater quality

The prime goal to monitor groundwater quality is to detect any short or long term adverse impacts caused by SAT schemes. According to NRC (2008), an efficient monitoring program should involve a frequent sampling schedule at the start of reuse scheme to develop historical records about its hydraulic characteristics and water quality trends. Two sets of wells are essential for a groundwater monitoring program. One installed hydraulically up gradient of the recharge area to know the condition of the incoming groundwater and the other located down gradient to monitor its quality against any deterioration that may arise as a result of the SAT scheme (Metcalf et al., 2007). Unless otherwise specified by regulations and guidelines, the monitoring program should include; field parameters (i.e. groundwater depth, pH, temperature,

EC, TDS and DO), *coliforms* bacteria, chemical oxygen demand (COD), nutrients, major ions, heavy metals and organic micropollutants when necessary.

8.3.5 Development of model to predict contaminants removal during SAT

A model was developed to predict contaminants removal from PE, SE and tertiary effluent (TE) for a specified pre-treatment process followed by SAT. The specific contaminants that were considered included DOC, NH_4-N, NO_3-N, PO_4-P, bacteria and viruses. While bacteria strains considered in the model include *total coliforms, faecal coliforms, cryptosporidium* and *giardia cysts*; viruses studied include polio virus, MS2 and PRD1.

8.3.5.1 Pre-treatment of wastewater effluent prior to SAT

Pre-treatment technologies considered were settling ponds (SP), coagulation and settling (CG), coagulation followed by rapid sand filtration (CG-RSF), disinfection (DN), granular activated carbon (GAC), micro-filtration (MF) and ultra-filtration (UF). Findings from the literature collected (35 cases) on various pre-treatment of wastewater effluent prior to application to SAT are listed in Table 8.5.

Table 8.5 Typical removal efficiency for pre-treatment of wastewater effluent

Parameter	Remov.	Settling Ponds	Roughing Filter	Coagulation (Coag+sed)	Rapid sand filtration (RSF)	Disinfection	Granular activated carbon (GAC)	Micro-filtration (MF)	Ultra-filtration (UF)	Reverse osmosis (RO)
TSS	%	50-70[1,2]	60-90[3]	80-90[2]	70-80[2]	–	–	95-98[2]	96-99.9[4]	95-100[2]
DOC	%	1-9.3[5,6]	11.5[6]	25-50[7,6]	–	–	30-70[8]	45-65[2]	50-75[4]	85-95[2]
NH$_4$-N	%	–	–	–	–	–	–	5-15[2]	5-15[4]	90-98[2]
NO$_3$-N	%	–	–	–	–	–	–	0-2[2]	0-2[4]	65-85[2]
PO$_4$-P	%	–	–	75-95[2]	–	–	–	0-2[2]	0-2[4]	95-99[2]
Bacteria	log$_{10}$	0.1-0.3[2]	–	80-90[2]	0-1[3,4]	1-4 (Cl)[2] and 1-4 (UV)[2]	–	2-5[11]	3-6[7]	4-7[11]
Giardia	log$_{10}$	<1[2]	–	–	0-3[4,8,9]	4 (UV)[3]	–	2-6[4]	4.7-7[3]	>7[2]
Crypto	log$_{10}$	0.1-1[2]	–	1.5-3[1]	1-2[1,11]	3-4 (UV)[1,10]	0.5-1[1]	0-2[4]	4.4-7[1,3]	4-7[2]
Viruses	log$_{10}$	<0.1[2]	–	2.7-7[2]	2[8,9]	1-4 (Cl)[2] and 1-4 (UV)[2]	–	0-2[2]	2-7[3,4]	4-7[2,7]

Source: [1]Degrémont (2007); [2]Tchobanoglous et al. (2003); [3]Au (2004); [4]Metcalf et al.(2007); [5]Katsoyiannis and Samara (2007); [6]Sharma et al. (2011); [7]Matilainen et al. (2010); [8]Crittenden and Harza (2005); [9]Schippers (2012); [10]Hendricks (2006); [11]Davis (2010)

Based on review of a number of cases reported in the literature for laboratory, pilot and field SAT studies, a spreadsheet model that predicts contaminants removal was developed.

8.3.5.2 Model elements and validation

Results from field studies and laboratory-based soil columns conducted in different parts of the world were used to predict contaminants removal from three types of wastewater effluent with respect to travel distance. Table 8.6, Table 8.7, Table 8.8 and Table 8.9 report contaminants concentrations, travel time, travel distance, the cases reviewed and removal efficiencies in different SAT systems.

Table 8.6 Summary of DOC removal in SAT system with respect to travel time and distance

Type of WW effluent	Influent concentration (mg/L)	Travel time (days)	Travel distance (m)	Number of cases	Removal (%)	Ref.
PE	9.9-42.5	0.1-8	0-5	19	12-61.1	1-7
SE	4.9-19	1-180	0-100	34	15-94	8-20
TE	4.3-19.4	0-360	0-100	33	17-82	21-28

Source: 1. Abel et al. (2014); 2. Abel et al. (2013a); 3. Abel et al. (2013b); 4. Abel et al. (2012); 5. Aharoni et al. (2010); 6. Cha et al. (2005); 7. Cha et al. (2004); 8. Drewes and Jekel (1998); 9. Drewes and Jekel (1996); 10. Fox (2006); 11. Fox et al. (2001a); 12. Fox et al. (2001b); 13. Hussen (2009);14. Jarusutthirak et al. (2003); 15. Katukiza (2006); 16. Laws et al. (2011); 17. Pi and Wang (2006); 18.Quanrud et al. (1996a); 19. Quanrud et al. (1996b); 20. Quanrud et al. (2003b); 21. Rauch and Drewes (2004); 22. Rauch and Drewes (2005); 23. Reemtsma et al. (2000); 24. Rice and Bouwer (1984); 25. Westerhoff and Pinney (2000); 26. Wilson et al. (1995); 27. Yoo et al. (2006); 28. Zhao et al. (2007).

Table 8.7 Summary of NH_4-N and NO_3-N removal in SAT system with respect to travel distance

Param.	Type of WW effl.	Infl. Conc. (mg/L)	Travel time (days)	Travel distance (m)	No. of cases	Removal (%)	Ref.
NH_4-N	PE	24-38.1	0.1-8	0.1-5	16	25-99.5	1-7
	SE	4.9-28	0.3-290	0.25-100	40	10.2-95.8	8-15
	TE	0.47-8.5	0.4-60	1.2-29	11	72-100	16-17
NO_3-N	SE	0.1-30	0.3-76	0.25-100	19	13-90	18-19
	TE	3.9-5.8	0.4-197	1.8-29	11	7-22	15-17

Source: 1. Abel et al. (2014); 2. Abel et al. (2013a); 3. Abel et al. (2013b); 4. Abel et al. (2012); 5. Aharoni et al. (2010); 6. Bekele et al. (2011); 7. Castillo et al. (2001); 8. Cha et al. (2005); 9. Fox (2006); 10. Fox et al. (2001a); 11. Kanarek et al. (1993); 12. Laws et al. (2011); 13. Leach and Enfield (1983); 14. Miller et al. (2006); 15. Mottier et al. (2000); 16. Nema et al. (2001); 17. Reemtsma et al. (2000); 18. Rice and Bouwer (1984); 19. Viswanathan et al. (1999).

Table 8.8 Summary of PO$_4$-P removal in SAT system with respect to travel distance

Type of WW effl.	Infl. Conc. (mg/L)	Travel time (days)	Travel distance (m)	No. of cases	Removal (%)	Ref.
PE	3.5-7.2	0.1-15.7	0.1-25	12	4-90	1-4
SE	0.7-10	4.2-6	2.8-100	15	30-99.4	5-9
TE	0.2-6.2	9-197	1.8-29	3	37-80	10-11

Source: 1. Abel et al. (2012); 2. Aharoni et al. (2010); 3. Bekele et al. (2011); 4. Bouwer et al. (1980); 5. Crites (1985); 6. Idelovitch et al. (2003); 7. Lance and Gerba (1980); 8. Nema et al. (2001); 9. Reemtsma et al. (2000); 10. Rice and Bouwer (1984); 11. Viswanathan et al. (1999).

Table 8.9 Summary of bacteria and virus removal in SAT system with respect to travel distance

Param.	Type of WW effl.	Infl. Conc. (CFU/100 mL)	Travel time (days)	Travel distance (m)	No. of cases	Removal (%)	Ref.
Bacteria	PE	0.6×10^6-2×10^7	0.5-250	0.3-25	9	1.2-6.9	1-6
	SE	12-1×10^7	7-20	1-100	13	1.6-no detection	7-9
	TE	286-0.6×10^6	38.2	29	2	2.4-no detection	10
Viruses	PE	1.2×10^4	4.6	0.8	1	4	2
	SE	2.1×10^3-2.3×10^7	0.1-20	1-9	36	0.0-4	8-12
	TE	–	0.1-4.6	1.0-5	30	0.4-4	11

Source: 1. Abel et al. (2014); 2. Abel et al. (2012); 3. Bouwer et al. (1980); 4. Castillo et al. (2001); 5. Idelovitch and Michail (1984); 6. Lance and Gerba (1980); 7. Mottier et al. (2000); 8. Nema et al. (2001); 9. Powelson et al. (1993); 10. Quanrud et al. (2003a); 11. Rice and Bouwer (1984); 12. Viswanathan et al. (1999).

For water quality prediction, the model requires four main input parameters namely (i) type of wastewater effluent, (ii) influent wastewater quality parameters (iii) type of pre-treatment technology and (iv) travel distance to the point of recovery relative to SAT infiltration basin. The type of wastewater effluent and type of pre-treatment can be selected from a dropdown list provided in the model. The horizontal (travel) distance cell allows the user to input a value from 0 to 100 m, which is translated by the model into a range of travel distances and displayed below the input cell as distance range. The range is displayed because the data used in the model related to SAT removal efficiencies is grouped based on ranges of travel distances. These ranges are 0-5 m, 5.1-10 m, 10.1-20 m and 20.1-100 m. The specific contaminants accepted by the model are DOC, NH$_4$-N, NO$_3$-N, PO$_4$-P, bacteria and viruses. Most data collected for SAT systems that receive PE were laboratory-scale setups with travel distance in the range 0-5 m. Likewise, travel distance ranged from 0-5 m for virus removal. Typical computation worksheet presenting input parameters and predicted removal efficiencies are presented in Figure 8.2 and Figure 8.3.

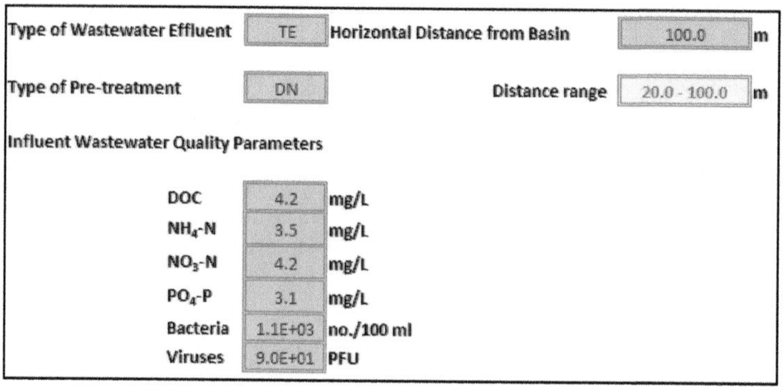

| Type of Wastewater Effluent | TE | Horizontal Distance from Basin | 100.0 | m |
| Type of Pre-treatment | DN | Distance range | 20.0 - 100.0 | m |

Influent Wastewater Quality Parameters

DOC	4.2	mg/L
NH_4-N	3.5	mg/L
NO_3-N	4.2	mg/L
PO_4-P	3.1	mg/L
Bacteria	1.1E+03	no./100 ml
Viruses	9.0E+01	PFU

Figure 8.2 Snapshot of the water quality prediction model showing input data

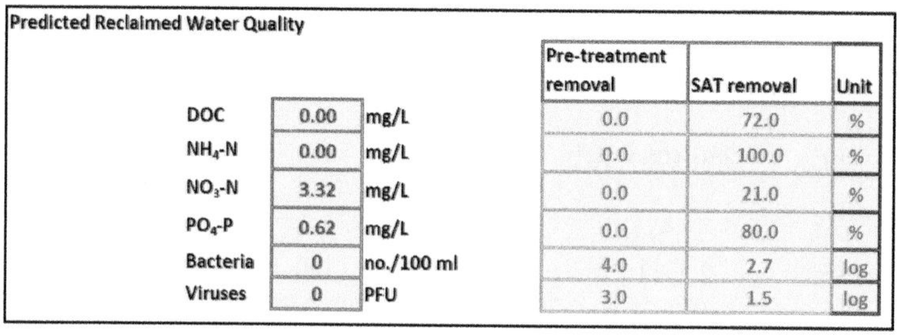

Predicted Reclaimed Water Quality

			Pre-treatment removal	SAT removal	Unit
DOC	0.00	mg/L	0.0	72.0	%
NH_4-N	0.00	mg/L	0.0	100.0	%
NO_3-N	3.32	mg/L	0.0	21.0	%
PO_4-P	0.62	mg/L	0.0	80.0	%
Bacteria	0	no./100 ml	4.0	2.7	log
Viruses	0	PFU	3.0	1.5	log

Figure 8.3 Snapshot of the water quality prediction model showing output data

A set of data obtained from the literature for PE, SE and TE was used to validate the model. 15 different concentrations PE, SE and TE falling within the ranges stated in Tables 8.4 - 8.6 were used in the model for various travel distance range. Based on these concentrations, the model predicted removal of the contaminants as a function in travel distance as shown in Figure 8.4, Figure 8.5 and Figure 8.6.

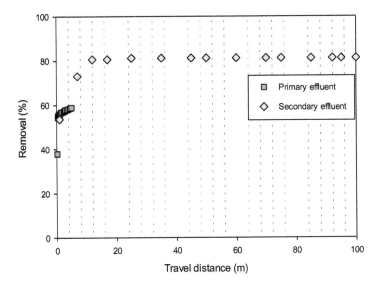

Figure 8.4 Plot of DOC removal with respective to travel distance in SAT
Source (Malolo, 2011; Buçpapaj, 2011)

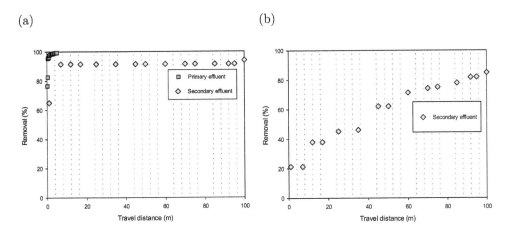

Figure 8.5 Plots of (a) NH_4-N and (b) NO_3-N removal with respective to travel distance in SAT
Source (Malolo, 2011; Buçpapaj, 2011)

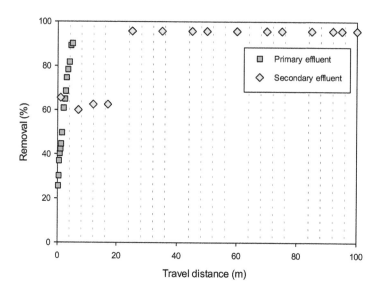

Figure 8.6 Plot of PO$_4$-P removal with respective to travel distance in SAT
Source (Malolo, 2011; Buçpapaj, 2011)

8.3.5.3 Assumptions and limitations of the model

Several assumptions were made while preparing the tools and water quality prediction model. On the other hand, the model had limitations.

a- Assumptions
- The tools were developed based on the assumption that they can serve as a detailed guide to planning, designing and post design activities of a SAT scheme.
- The pre-screening tool is built assuming that SAT amongst other treatment technologies would have already been chosen and therefore, it does not consider evaluation of other treatment technologies.
- A WWTP exists and provides a certain level of treatment for wastewater, while SAT is considered as an added treatment stage or a polishing stage of the effluent prior to reuse. Furthermore, SAT site is identified and selected taking the WWTP location as the reference point.
- The site selection tool assumes that topographical, land use, geological and hydro-geological maps exist and can be used to extract data that will be used in the tool.
- The data used in the model for contaminats removal are mostly based on vertical and horizontal removals measured in laboratory (soil column) experiments, pilot studies through piezometers set below SAT basins and few field production wells. In the case of soil columns, it is assumed that horizontal removals are negligible. This implies that values obtained from the model are rather an underestimation of the real life values expected on site.

- The predicted contaminants removals were based on soil types commonly used in the fields and laboratory experiments from which the data were collected (namely, fine sand, loamy sand and sandy loam).

  b- Limitations

- Different types of SAT systems have been discussed (i.e. infiltration basin, vadose zone wells, and direct injection wells). However, the guidelines and tools developed are specifically meant for SAT by infiltration basin. This is due to the wide scope of each SAT system that requires time to jointly cover all the SAT systems.

- The data used for contaminants removal efficiencies are based on laboratory (soil column) experiments and handful SAT site data that in most cases do not represent the actual heterogeneity of SAT sites. The field data as well are from SAT basin operated under different operational and climatic conditions. Therefore, the model can only provide broad range of values that can only serve as an estimation of the expected contaminants removal through SAT.

- Except for the travel distance range of 20-100 m used in the model for water quality prediction, all the other ranges do not take into consideration influence of dilution with local groundwater. This is because all data collected for the other travel distance ranges represent water quality changes within the vadose zone only.

- For more representative prediction and suitability for SAT sites (case by case), another input window should be added to indicate soil type at a particular site.

- The model does not predict removal of organic micropollutants and more effort is needed to include these compounds in the model.

8.3.6 Practical use of tools and model developed

The tools and model developed during this research are to be used independently relative to the stage of the SAT scheme and the intended user. The pre-screening tool is intended for decision makers and planners and is to be used at the planning stage of SAT project. The user should have an understanding of the relevant institutions and their setup, water reuse regulations and general information on the hydrogeology of the region. The site selection tool can be used when land use, topographical, geological, hydrological maps and some general details of WWTP are available. Once the data are studied, the tool can be filled to assess the most suitable site. On the other hand, site investigation tool can be used to identify the field works required and to make an estimate of the costs. The design tool details the steps involved in the design process. Furthermore, the water quality prediction model developed can be used to get an estimation of contaminants removal based on wastewater effluent characteristics, pre-treatment and distance between infiltration basin and point of recovery. Table 8.10 summarizes the tools developed and their intended users.

Table 8.10 Summary of tools developed, intended users and input data

Tool	Input data	Intended users
SAT pre-screening	Institutional, economic, regulatory, socio-political and technical aspects data	Decision makers and planners
Site identification and selection	Hydrogeological, land use, land grade and site susceptibility to flooding	Engineers and planners
Site investigation	Results from field investigation on test pits, boreholes, soil, infiltration tests, groundwater quality	Engineers and laboratory technicians
SAT scheme design	Type of wastewater effluent, volume, length of wetting/drying, BOD and nitrogen loading rate, land topography and local climate	Engineers
SAT operation and maintenance	Wetting/drying and periodic cleaning programs	Engineers and operators
Reclaimed water quality prediction model	Type of wastewater effluent and characteristics, type of pre-treatment and horizontal distance from infiltration basin,	Engineers and operators

In general, the tools provide detailed requirements of a SAT scheme and can be therefore used for cost estimation of SAT scheme implementation. These tools along with the water quality prediction model provide a base for comparison of SAT with other wastewater reuse systems. It is expected that the use of tools and model developed will help to promote SAT technology in developing countries.

8.4 CONCLUSIONS

The tool developed for SAT pre-screening showed that institutional, economical and legal aspect are much critical than regulatory and socio-political aspects during the SAT planning stages. Of most importance are public involvement, public perception and cost of land.

SAT site identification is governed by availability of land area, depth of vadose zone and proximity of SAT site to WWTP. However, field investigation to physically assess and explore the site by conducting test pits, infiltration tests, groundwater wells and boreholes site is of paramount importance. This if followed by laboratory analysis of the samples collected during the field work.

Results obtained from water quality prediction model for DOC, NH_4-N, NO_3-N, PO_4-P, bacteria and viruses revealed that efficiency of SAT to remove these contaminants is dependent on the type of wastewater effluent, pre-treatment provided and travel distance.

The tools developed under this chapter could be used by different users during preliminary, design and operation stages of SAT scheme. However, each site has its own characteristics and these tools serve as general guide. Factors such as soil type, availability of reuse regulations, land use and availability of restricting materials in the vadose zone should be carefully checked and assessed before choosing SAT for water reclamation.

8.5 REFERENCES

Abel, C. D., Sharma, S. K., Mersha, S. A. and Kennedy, M. D. (2014). Influence of intermittent infiltration of primary effluent on removal of suspended solids, bulk organic matter, nitrogen and pathogens indicators in a simulated managed aquifer recharge system. *Ecological Engineering,* **64**, 100-107.

Abel, C. D. T., Sharma, S. K., Buçpapaj, E. and Kennedy, M. D. (2013a). Impact of hydraulic loading rate and media type on removal of bulk organic matter and nitrogen from primary ef uent in a laboratory-scale soil aquifer treatment. *Water Science and Technology,* **68**(1), 217-226.

Abel, C. D. T., Sharma, S. K., Maeng, S. K., Magic-Knezev, A., Kennedy, M. D. and Amy, G. L. (2013b). Fate of Bulk Organic Matter, Nitrogen, and Pharmaceutically Active Compounds in Batch Experiments Simulating Soil Aquifer Treatment (SAT) Using Primary Effluent. *Water, Air, and Soil Pollution,* **224**(7), 1-12.

Abel, C. D. T., Sharma, S. K., Malolo, Y. N., Maeng, S. K., Kennedy, M. D. and Amy, G. L. (2012). Attenuation of Bulk Organic Matter, Nutrients (N and P), and Pathogen Indicators During Soil Passage: Effect of Temperature and Redox Conditions in Simulated Soil Aquifer Treatment (SAT). *Water, Air, and Soil Pollution,* **223**, 5205-5220.

Aharoni, A., Guttman, Y., Tal, N., Kreitzer, T. and Cikurel, H. (2010). SWITCH project Tel-Aviv Demo City, Mekorot's case: hybrid natural and membranal processes to up-grade effluent quality. *Reviews in Environmental Science and Biotechnology,* **9**(3), 193-198.

Asano, T. and Cotruvo, J. A. (2004). Groundwater recharge with reclaimed municipal wastewater: health and regulatory considerations. *Water Research,* **38**(8), 1941-1951.

Assaf, S. A. and Al-Hejji, S. (2006). Causes of delay in large construction projects. *International Journal of Project Management,* **24**(4), 349-357.

Au, K.-K. (2004). *Water Treatment and Pathogen Control: Process Efficiency in Achieving Safe Drinking-water.* IWA Publishing.

Bekele, E., Toze, S., Patterson, B. and Higginson, S. (2011). Managed aquifer recharge of treated wastewater: Water quality changes resulting from infiltration through the vadose zone. *Water Research,* **45**(17), 5764-5772.

Bouwer, H., Rice, R., Lance, J. and Gilbert, R. (1980). Rapid-infiltration research at Flushing Meadows project, Arizona. *Journal of Water Pollution Control Federation,* 2457-2470.

Buçpapaj, E. (2011). Effect of Soil Type and Hydraulic Regime on Removal of Bulk Organic Matter and Nitrogen During Soil Passage. UNESCO-IHE MSc thesis MWI 2011-04, Delft, the Netherlands.

Castillo, G., Mena, M., Dibarrart, F. and Honeyman, G. (2001). Water quality improvement of treated wastewater by intermittent soil percolation. *Water Science and Technology*, **43**(12), 187-190.

Cha, W., Choi, H., Kim, J. and Cho, J. (2005). Water quality dependence on the depth of the vadose zone in SAT-simulated soil columns. *Water Science and Technology: Water Supply*, **5**(1), 17-24.

Cha, W., Choi, H., Kim, J. and Kim, I. (2004). Evaluation of wastewater effluents for soil aquifer treatment in South Korea. *Water Science and Technology*, **50**(2), 315-322.

Crites, R., Reed, S. and Bastian, R. (2000). *Land Treatment Systems for Municipal and Industrial Wastes*. McGraw-Hill Professional.

Crites, R. W. (1985). Nitrogen removal in rapid infiltration systems. *Journal of Environmental Engineering*, **111**(6), 865-873.

Crites, R. W., Reed, S. C. and Middlebrooks, E. J. (2006). *Natural Wastewater Treatment Systems*. CRC Press, Boca Raton, Florida, USA, pp 413-426.

Crittenden, J. and Harza, M. W. (2005). *Water Treatment: Principles and Design*. John Wiley & Sons, New Jersy, USA.

Crook, J. and Surampalli, R. Y. (1996). Water reclamation and reuse criteria in the US. *Water Science and Technology*, **33**(10), 451-462.

Davis, M. (2010). *Water and Wastewater Engineering*. McGraw-Hill Science/Engineering/Math. ISBN: 0073397865.

Degrémont. (2007). Water Treatment Handbook. Lavoisier Publishing. ISBN-10: 2743009705.

Drewes, J. and Jekel, M. (1998). Behavior of DOC and AOX using advanced treated wastewater for groundwater recharge. *Water Research*, **32**(10), 3125-3133.

Drewes, J. E. (2009). Ground water replenishment with recycled water—water quality improvements during managed aquifer recharge. *Ground Water*, **47**(4), 502-505.

Drewes, J. E. and Jekel, M. (1996). Simulation of groundwater recharge with advanced treated wastewater. *Water Science and Technology*, **33**(10), 409-418.

Fox, P. (2006). *Advances in Soil Aquifer Treatment Research for Sustainable Water Reuse*. American Water Works Association (AWWA) Research Foundation and AWWA. ISBN: 1583214372.

Fox, P., Houston, S. and Westerhoff, P. (2001a). *Soil Aquifer Treatment for Sustainable Water Reuse*. American Water Works Association, Denver, Clorado, USA.

Fox, P., Narayanaswamy, K., Genz, A. and Drewes, J. (2001b). Water quality transformations during soil aquifer treatment at the Mesa Northwest Water Reclamation Plant, USA. *Water Science and Technology*, 343-350.

Hartley, T. W. (2006). Public perception and participation in water reuse. *Desalination*, **187**(1-3), 115-126.

Hendricks, D. W. (2006). *Water Treatment Unit Processes: Physical and Chemical*. CRC Press.

Huertas, E., Salgot, M., Hollender, J., Weber, S., Dott, W., Khan, S., Schäfer, A., Messalem, R., Bis, B., Aharoni, A. and Chikurel, H. (2008). Key objectives for water reuse concepts. *Desalination,* **218**(1-3), 120-131.

Hussen, M., A. (2009). Advanced Primary Pre-treatment for Soil Aquifer Treatment (SAT). UNESCO-IHE MSc thesis MWI 2009-10, Delft, the Netherlands.

Idelovitch, E., Icekson-Tal, N., Avraham, O. and Michail, M. (2003). The long-term performance of Soil Aquifer Treatment(SAT) for effluent reuse. *Water Science and Technology: Water Supply,* **3**(4), 239-246.

Idelovitch, E. and Michail, M. (1984). Soil-aquifer treatment: a new approach to an old method of wastewater reuse. *Journal of Water Pollution Control Federation,* **56**(8), 936-943.

Jarusutthirak, C., Amy, G. and Foss, D. (2003). Potable reuse of wastewater effluent through an integrated soil aquifer treatment (SAT)-membrane system. *Water Supply,* **3**(3), 25-33.

Kanarek, A., Aharoni, A. and Michail, M. (1993). Municipal wastewater reuse via soil aquifer treatment for non-potable purposes. *Water Science and Technology,* **27**(7-8), 53-61.

Katsoyiannis, A. and Samara, C. (2007). The fate of dissolved organic carbon (DOC) in the wastewater treatment process and its importance in the removal of wastewater contaminants. *Environmental Science and Pollution Research-International,* **14**(5), 284-292.

Katukiza, A. (2006). Effect of Wastewater Quality and Process Parameters on Removal of Organic Matter During Soil Aquifer Treatment, UNESCO-IHE MSc thesis. MWI 2006-15. Delft, the Netherlands.

Lance, J. and Gerba, C. (1980). Poliovirus movement during high rate land filtration of sewage water. *Journal of Environmental Quality,* **9**(1), 31-34.

Laws, B. V., Dickenson, E. R. V., Johnson, T. A., Snyder, S. A. and Drewes, J. E. (2011). Attenuation of contaminants of emerging concern during surface-spreading aquifer recharge. *Science of the Total Environment,* **409**(6), 1087-1094.

Le-Hoai, L., Dai Lee, Y. and Lee, J. Y. (2008). Delay and cost overruns in Vietnam large construction projects: a comparison with other selected countries. *KSCE Journal of Civil Engineering,* **12**(6), 367-377.

Leach, L. E. and Enfield, C. G. (1983). Nitrogen control in domestic wastewater rapid infiltration systems. *Journal of Water Pollution Control Federation,* 1150-1157.

Malolo, Y. (2011). Effect of Temperature and Redoc Conditions on Removal of Contaminants During Soil Aquifer treatment. UNESCO-IHE MSc thesis MWI 2011-09, Delft, the Netherlands.

Matilainen, A., Vepsäläinen, M. and Sillanpää, M. (2010). Natural organic matter removal by coagulation during drinking water treatment: A review. *Advances in Colloid and Interface Science,* **159**(2), 189-197.

Metcalf, E., Asano, T., Burton, F., Leverenz, H., Tsuchihashi, R. and Tchobanoglous, G. (2007). *Water Reuse: Issues, Technologies, and Applications,* Mc-Graw Hill. NewYork, USA.

Miller, J. H., Ela, W. P., Lansey, K. E., Chipello, P. L. and Arnold, R. G. (2006). Nitrogen Transformations During Soil–Aquifer Treatment of Wastewater Effluent—Oxygen Effects in Field Studies. *Journal of Environmental Engineering*, **132**, 1298.

Miotli ski, K., Barry, K., Dillon, P. and Breton, M. (2010). Alice Springs SAT Project Hydrological and Water Quality Monitoring Report 2008-2009. CSIRO. 1835-095X.

Mottier, V., Brissaud, F., Nieto, P. and Alamy, Z. (2000). Wastewater treatment by infiltration percolation: a case study. *Water Science and Technology*, **41**(1), 77-84.

Nema, P., Ojha, C., Kumar, A. and Khanna, P. (2001). Techno-economic evaluation of soil-aquifer treatment using primary effluent at Ahmedabad, India. *Water Research*, **35**(9), 2179-2190.

Pi, Y.-z. and Wang, J.-l. (2006). A field study of advanced municipal wastewater treatment technology for artificial groundwater recharge. *Journal of Environmental Sciences*, **18**(6), 1056-1060.

Powelson, D., Gerba, C. and Yahya, M. (1993). Virus transport and removal in wastewater during aquifer recharge. *Water Research*, **27**(4), 583-590.

Quanrud, D., Arnold, R., Wilson, L. and Conklin, M. (1996a). Effect of soil type on water quality improvement during soil aquifer treatment. *Water Science and Technology*, **33**(10), 419-432.

Quanrud, D., Arnold, R., Wilson, L., Gordon, H., Graham, D. and Amy, G. (1996b). Fate of organics during column studies of soil aquifer treatment. *Journal of Environmental Engineering*, **133**(4), 314-321.

Quanrud, D., Carroll, S., Gerba, C. and Arnold, R. (2003). Virus removal during simulated soil-aquifer treatment. *Water Research*, **37**(4), 753-762.

Rauch, T. and Drewes, J. (2004). Assessing the removal potential of soil-aquifer treatment systems for bulk organic matter. *Water Science and Technology*, **50**(2), 245-253.

Rauch, T. and Drewes, J. (2005). Quantifying biological organic carbon removal in groundwater recharge systems. *Journal of Environmental Engineering*, **131**(6), 909-923.

Reed, S. C., Crites, R. and Wallace, A. (1985). Problems with Rapid Infiltration: A Post Mortem Analysis. *Journal of Water Pollution Control Federation*, 854-858.

Reemtsma, T., Gnir , R. and Jekel, M. (2000). Infiltration of combined sewer overflow and tertiary municipal wastewater: an integrated laboratory and field study on nutrients and dissolved organics. *Water Research*, **34**(4), 1179-1186.

Rice, R. and Bouwer, H. (1984). Soil-aquifer treatment using primary effluent. *Journal of Water Pollution Control Federation*, 84-88.

Schippers, J. C. (2012). Disinfection Process. MSc Lecture Notes. UNESCO-IHE, Delft, the Netherlands.

Sharma, S. K., Hussen, M. and Amy, G. L. (2011). Soil aquifer treatment using advanced primary effluent. *Water Science and Technology*, **64**(3), 640-646.

Tchobanoglous, G., Burton, F. L. and Stensel, H. D. (2003). *Wastewater Engineering: Treatment and Reuse. 4th ed.* Metcalf and Eddy. McGraw-Hill Company. ISBN 0-07-112250-8.

USEPA. (2012). Guidelines for Water Reuse. *In Guideline for water reuse,* CDM Smith. Washington, USA.

USEPA. (2006). Process Design Manual: Land Treatment of Municipal Wastewater Effluents. EPA/625/R-06/016, US Environmental Protection Agency, Office of Research and Development, Cincinnati, Ohio, USA.

Viswanathan, M., Al Senafy, M., Rashid, T., Al-Awadi, E. and Al-Fahad, K. (1999). Improvement of tertiary wastewater quality by soil aquifer treatment. *Water Science and Technology,* **40**(7), 159-163.

Westerhoff, P. and Pinney, M. (2000). Dissolved organic carbon transformations during laboratory-scale groundwater recharge using lagoon-treated wastewater. *Waste Management,* **20**(1), 75-83.

Wilson, L., Amy, G., Gerba, C., Gordon, H., Johnson, B. and Miller, J. (1995). Water quality changes during soil aquifer treatment of tertiary effluent. *Water Environment Research,* **67**(3), 371-376.

Yoo, H. H., Miller, J. H., Lansey, K. and Reinhard, M. (2006). EDTA, NTA, alkylphenol ethoxylate and DOC attenuation during soil aquifer treatment. *Journal of Environmental Engineering,* 132, 674.

Zhao, Q., Xue, S., You, S. and Wang, L. (2007). Removal and transformation of organic matter during soil aquifer treatment. *Journal of Zhejiang University Science A,* **8**(5), 712-718.

CHAPTER 9

SUMMARY AND CONCLUSIONS

9.1   SOIL AQUIFER TREATMENT USING PRIMARY EFFLUENT:
POTENTIAL AND CHALLENGES

Raw wastewater is discharged untreated to surface water in the vast majority of developing countries due to lack of investment to construct and operate WWTPs that are capable of treating wastewater to secondary and tertiary effluent levels. Disposal of untreated wastewater poses adverse health impacts on communities using river water for drinking purposes since drinking water treatment plants either do not exist or existing plants do not meet water quality requirements. Besides, application of raw wastewater in agricultural irrigation to produce crops, direct contact between farmers and this untreated wastewater causes severe risks to human health. To overcome these health risks, a cost-effective, low-tech and robust wastewater treatment and reuse system that requires marginal amounts of energy and chemicals is needed. Land application systems (i.e. soil aquifer treatment) are among the most attractive systems that could be used for this purpose.

Soil aquifer treatment (SAT) is a natural system that has been successfully employed worldwide for wastewater treatment and reuse. It utilizes soil strata to produce water of acceptable quality for intended use. Treatment benefits during SAT are achieved during infiltration of wastewater effluent initially through the unsaturated zone and eventually in the saturated zone in the aquifer where it mixes with the native groundwater before it is recovered via a production well for reuse.

SAT has the potential to augment existing water resources, divert wastewater from receiving water bodies used as drinking water sources and provide a psychological barrier in communities where social and religious taboos inhibit any treatment option that does not include utilization of land treatment. Furthermore, SAT is a resilient treatment option that covers a wide range of wastewater effluents (primary, secondary and tertiary). The use of primary effluent in SAT is an attractive option for developing countries and provides economical benefits since wastewater treatment to this level prior to application to SAT does not require a sizable investment. Besides, SAT system does not require extensive use of energy and chemicals. However, SAT systems are site-specific, lack reliable tools to transfer experience, require sizable land area and lack a framework for feasibility, design and operation. Another constraining factor for SAT using PE is the low infiltration rate that causes water retention in infiltration basins and subsequent high evaporation potential, especially in warm dry climates.

Even though SAT has been used for wastewater treatment and reuse for decades, there are knowledge gaps considering the establishment of new SAT schemes, especially where the technology could be used to its maximum potential in developing countries. Furthermore, process conditions and removal of different contaminants are not fully understood and require more investigation. The aim of this research work was to close this gap and get a better understanding of the effect of soil type, hydraulic loading rate, change in temperature and biological activity on the removal of suspended solids, bulk organic matter, nitrogen, phosphorus, pharmaceutically active compounds and pathogen indicators during SAT. Tools and a water quality prediction model were also developed to conduct SAT feasibility studies, design, operation and estimate potential contaminant removal in SAT schemes.

9.2 EFFECT OF PRE-TREATMENT OF PRIMARY EFFLUENT USING ALUMINUM SULFATE AND IRON CHLORIDE ON REMOVAL OF SUSPENDED SOLIDS, BULK ORGANIC MATTER, NUTRIENTS AND PATHOGENS INDICATORS

The rapid clogging of infiltration basins due to the relatively high concentration of suspended solids (SS) is one of the main limitations of SAT systems using PE. The effect of pre-treatment of PE from domestic wastewater prior to application to SAT was investigated using optimum dosages of both aluminium sulfate and iron chloride as coagulants. A 4.2 m high laboratory-scale soil column was constructed and used to simulate a SAT system. The results demonstrated similar overall removal of ~ 90% of SS for coagulated and non-coagulated SAT influent. On the other hand, coagulation of PE increased removal of dissolved organic carbon (DOC) from 16-22% through infiltration only to ~ 70% achieved through infiltration of coagulated PE using both coagulants. Likewise, removal of phosphorus (PO_4-P) increased from as low as ~ 30% during infiltration of non-coagulated PE to >98% as a result of PE coagulation prior to infiltration. *E.coli* and *total coliforms* removal increased from 2.5 $\log_{10}$ units for non-coagulated PE to >4 $\log_{10}$ units for coagulated PE for both coagulants. These findings clearly suggest that pre-treatment of PE using aluminium sulfate and iron chloride reduces the contaminants analyzed and improves both water quality and operation of SAT spreading basins. Both coagulants can equally reduce land area and minimizes post-treatment requirements for reclaimed water. Moreover, locally available organic coagulants (i.e. *moringa oleifera*, chitosan and okra) could be used to pre-treat PE prior to application to SAT in some developing countries where chemical coagulants are unaffordable.

9.3   IMPACT OF HYDRAULIC LOADING RATE AND SOIL TYPE ON REMOVAL OF BULK ORGANIC MATTER AND NITROGEN FROM PRIMARY EFFLUENT IN LABORATORY-SCALE SOIL AQUIFER TREATMENT SYSTEM

Hydraulic loading rate (HLR) and soil type play a pivotal role in successful operation of SAT scheme. This study explored the effect of HLR and soil type on removal of bulk organic matter and nitrogen from PE (90-95% domestic, 5-10 industrial wastewater) using a 5 m laboratory-scale column. The HLR was varied from 0.625 to 1.25 m/day and the effectiveness of two different types of sands (silica sand and dune sand) was examined.

Results from the experiments revealed respective DOC removals of 47.5±1.2 and 45.1±1.2% in silica sand columns operated at HLRs of 0.625 and 1.25 m/d implying that DOC removal was not significantly affected by the change in HLR. However, the dune sand column operated at 1.25 m/d exhibited DOC removal of 57.3±7.6%, implying that soil type used has more significant effect on DOC removal efficiency than HLR. Furthermore, Ammonium nitrogen (NH_4-N) removals of 74.5±18.0 and 39.1±4.3% in silica sand column operated at 0.625 and 1.25 m/d, respectively. NH_4-N removal of 49.2±5.2% was achieved in dune sand operated at HLR of 1.25 m/d. These results imply that ammonium removal during soil passage is influenced by both HLR as well as soil type.

Results of this study show that while DOC removal from PE in SAT system is not dependent on soil type or HLR, DOC removal in two different soils is likely to be higher in the one with finer particles which provides more surface area for biofilms to develop. Furthermore, the findings suggest that NH_4-N reduction in SAT is likely to be relatively higher when PE is applied at low HLR and soil with fine particles due to longer contact time and presence of adsorption binding sites. The practical implication of using such HLR and soil type is the need for much frequent drying and scraping of soil surface to remove the clogging layer.

9.4   INFLUENCE OF WETTING AND DRYING CYCLES ON REMOVAL OF SUSPENDED SOLIDS, BULK ORGANIC MATTER, NUTRIENTS AND PATHOGENS INDICATORS FROM PRIMARY EFFLUENT IN MANAGED AQUIFER RECHARGE

One of the fundamental operational aspects of SAT is alternate wetting and drying of the spreading basins to allow gaseous oxygen to penetrate into the vadose zone and aerate soil beneath the surface of the basin. Two laboratory-scale soil columns with 4.2 m height were fed with PE and operated at HLR of 0.625 and 1.25 m/d. Continuous wetting for 6.4 days was coupled with varying drying periods of 1, 3.2

and 6.4 days at HLR of 0.625 m/d while continuous wetting of 3.2 days was coupled with 1, 3.2 and 6.4 day drying periods.

Experimental results obtained, showed no positive correlation between DOC (57%) and SS (90%) removals with the length of drying period applied to the system. Nevertheless, NH_4-N, *E. coli* and *total coliforms* removal increased significantly with the length of the drying period.

The study demonstrated that SAT systems fed with PE and operated at HLR of 0.625 m/d could be operated at wetting/drying of 1:0.5, while a wetting/drying of 1:1 is applicable at SAT systems operated at HLR of 1.25 m/d.

9.5 EFFECT OF BIOLOGICAL ACTIVITY ON REMOVAL OF BULK ORGANIC MATTER, NITROGEN AND PHARMACEUTICALLY ACTIVE COMPOUNDS FROM PRIMARY EFFLUENT

Biodegradation, adsorption, chemical precipitation and ion-exchange are the main mechanisms of removal of different contaminants during soil passage. It is important to get some insight into the key removal mechanisms for different contaminants during soil passage so that the design and O&M of SAT systems could be optimized. Laboratory-scale batch reactors filled with silica sand, were periodically fed with PE and ripened for varying periods. While blank samples contained only PE, control reactors were filled with silica sand and PE at 5 day intervals. On the other hand, biologically active reactors were ripened for 240 days followed by suppression of biological activity in some of the reactors using different concentrations of a biocide (sodium azide).

The results obtained suggest that ripening of batch reactors and consequently acclimation of SAT systems plays a pivotal role in the removal of bulk organic matter, nitrogen and most PhACs of interest. While a ripening time of 5 days resulted in a DOC removal of 30.4±5.0 mg/L, a reactor that was operated for 240 days was capable of removing 75.3±2.4% of DOC content of PE with biologically suppressed reactors showing DOC removal between these levels. It was established that biological activity (measured as adenosine triphosphate (ATP)) correlated significantly with DOC removal. Similar trends were observed for NH_4-N where low removal (10.4±2.5%) was observed in blank reactors and almost complete reduction of NH_4-N was achieved in the reactors ripened for 240 days. It was also noticed that the reactors ripened for 240 days, with high biological activity showed improved removal of gemfibrozil, diclofenac and bezafibrate from <20% in reactors ripened for 5 days to >90%. Phenacetin, paracetamol, ibuprofen and caffeine were easily removed under various operating conditions. The obtained results imply that while new SAT systems would easily remove phenacetin, paracetamol and caffeine even during the first flooding after drying or scraping of the SAT infiltration basin, substantial removal of gemfibrozil, diclofenac, pentoxifylline and bezafibrate from PE would require high biological activity after long ripening periods (i.e. 240 days). The results also show that fully ripened SAT system (i.e. >240 days) can substantially (70.1±2.3%) remove clofibric acid.

9.6 EFFECTS OF TEMPERATURE AND REDOX CONDITIONS ON ATTENUATION OF BULK ORGANIC MATTER, NITROGEN, PHOSPHORUS AND PATHOGENS INDICATORS DURING MANAGED AQUIFER RECHARGE

The performance of SAT mainly depends on wastewater effluent quality, hydrogeological conditions at site and process conditions applied. Temperature and redox conditions are reported as key parameters influencing the removal of contaminants during soil passage. Effect of temperature and redox conditions on removal of bulk organic matter, nutrients (N and P) and pathogens indicators from PE was explored using laboratory-scale 0.3 m columns and batch reactors. Experimental results revealed that DOC removal increased by 5% for each 5°C increase in temperature in the range 15 to 25°C. DOC removal increased from as low as 17.1±6.4% at 5°C to as high as 54.4±0.4% at 25°C. On the other hand, DOC removal (46.4±2.0%) under aerobic operating conditions was 15% higher than the removal achieved under anoxic operating conditions.

NH_4-N removal of 89.7 - 99% was attained at 15 - 25°C through nitrification leading to notable increase in concentration of NO_3-N. Nevertheless, NH_4-N removal decreased significantly at 5°C to 8.8%. PO_4-P removal increased progressively with increase in temperature suggesting that low water viscosity at high temperature increased diffusion to adsorption sites.

Results obtained in this study demonstrated that the efficiency of SAT system to remove bulk organic matter, nitrogen and pathogens indicators improved significantly at high temperature. Likewise, higher contaminant removal was achieved with aerobic operating conditions compared to anoxic conditions, implying that aeration of PE prior to infiltration and application of wetting and drying cycles can improve the performance of SAT.

9.7 FRAMEWORK FOR SITE SELECTION, DESIGN, OPERATION AND MAINTENANCE OF SOIL AQUIFER TREATMENT (SAT) SYSTEM

SAT systems have been successfully used for water reclamation and reuse for several decades in different parts of the world. However, lack of detailed guidelines and a framework that provides insightful guidance for new SAT proponents to develop new schemes based on a step-by-step tool for planners, engineers and decision makers is one of the challenges faced during development of new SAT schemes.

In this study, a framework for SAT pre-feasibility was developed along with tools for site identification, selection and investigation, SAT site design, operation and monitoring. Furthermore, a spreadsheet model that predicts potential removal of selected contaminants was also developed. Some of the methods used included extensive literature review for data collection, survey distribution, model development, validation and verification using data from SAT field sites and laboratory-scale soil columns.

The findings from the study revealed that factors like public involvement, land cost, potential location of the SAT scheme relative to the wastewater treatment plant (WWTP), establishment of institutional framework and the reuse market should be critically analyzed and considered during pre-feasibility stages. Likewise, the development of a site selection tool was based on three site specific factors (physical, hydrogeological, land use and economical). This tool is meant for the planning stage to choose the most suitable site from a number of sites. A summary of site investigation and laboratory analysis tests required at new SAT schemes is provided. A tool for SAT system design including detailed considerations, parameters and steps was also developed. The O&M and monitoring tool was developed for post-design stage. This tool elaborates the requirements of a SAT scheme during operation. Furthermore, an Excel-based water quality prediction model was also developed to predict organic matter, nitrogen, phosphorus, bacteria and virus removal based on travel distance through SAT in combination with selected pre-treatment technologies

9.8 PRACTICAL IMPLICATIONS OF THE FINDINGS AND PROSPECTS FOR FURTHER RESEARCH

The outcomes of this study clearly demonstrate that SAT using PE is an effective means of wastewater treatment and reuse, especially in countries where investment costs as well as high O&M requirements hinder the construction of conventional WWTPs.

Based on the findings of this study, coagulation of PE is a viable option that removes suspended solids and consequently reduces the clogging potential and area requirements as infiltration rates are improved. The operational significance of the study suggests that the operation of SAT sites receiving PE in winter with water temperatures below 5°C will have a negative impact on the performance of these sites with regard to removal of bulk organic matter, nutrients and pathogens indicators. On the other hand, performance of these sites at temperature higher than 15°C in summer will increase removal of the above mentioned contaminants. Furthermore, aeration during PE application method to infiltration basins (DO >5 mg O_2/L) to PE will yield more removal of bulk organic matter, nutrients and pathogens indicators as compared with application of unaerated PE. It is also of paramount importance to note that long drying periods for infiltration basins will result in better removal of nitrogen, E. Coli and total coliforms compared to short drying periods. High (>90%) removal of some PhACs from PE can be achieved at relatively old SAT sites through biological mechanisms compared to relatively new SAT sites. The developed framework and tools for scheme feasibility, design and operation of SAT can be effectively used by planners, engineers and operators in regions with water scarcity that would like to incorporate SAT in their integrated water management plan.

Further research is required to investigate the effect of temperature at 30°C on the removal of bulk organic matter, nutrients and pathogens indicators from PE. Effect of temperature and redox conditions on the fate of PhACs and endocrine disrupting compounds (EDCs) is another potential research field. Furthermore, the use of locally available materials (i.e. *moringa oleifera*, chitosan, etc.) as coagulants for treatment of PE where metal based, inorganic coagulants are not affordable requires a thorough investigation. Lastly, further research is required to predict the removal of OMPs in SAT at different travel distances and pre-treatment options.

SAMENVATTING

De lozing van onbehandeld of onvoldoende behandeld afvalwater in meren, beken en op grondgebied stijgt globaal met duizelingwekkende volumes, met name in ontwikkelingslanden, als gevolg van de snelle bevolkingsgroei, verstedelijking en het gebrek aan investeringen om conventionele rioolwaterzuiveringsinstallaties (RWZI's) te bouwen, te exploiteren en te onderhouden. Bovendien is de meerderheid van de bestaande WWPTs (als ze er al zijn) verouderd en overbelast omdat ze zijn ontworpen om slechts kleine fractie van de huidige populatie te bedienen. Aan de andere kant is er toenemende waterschaarste in verschillende delen van de wereld en een sterke concurrentie over water tussen de verschillende sectoren. Als gevolg hiervan is de ontwikkeling en implementatie van rendabele en milieuvriendelijke verwerkingstechnologieën met een lage energie- en chemische voetafdruk gewenst om vervuiling van het oppervlaktewater te verminderen, en om een effectief integraal waterbeheer aan te kunnen bieden middels hergebruik van water. Gecontroleerde toepassingen van afvalwater op het land, zoals bodeminfiltratie van voorgezuiverd afvalwater (soil aquifer treatment, SAT) hebben de potentie om afvalwatereffluent te behandelen voor later hergebruik.

SAT is een geozuiveringssysteem waarbij afvalwatereffluent wordt geïnfiltreerd in bodemlagen om de waterkwaliteit te verbeteren, waarbij fysische, chemische en biologische processen een rol spelen. Het eerste gedeelte van de behandeling vindt plaats tijdens verticale infiltratie van afvalwatereffluent door de onverzadigde bodemzone; verdere behandeling gebeurt tijdens de horizontale beweging door de verzadigde zone voordat het weer geabstraheerd wordt uit een waterbron. Hoewel SAT op verschillende locaties over de hele wereld gebruikt wordt voor de verdere behandeling en hergebruik van afvalwatereffluent, is de meeste ervaring plaatsgebonden en er zijn geen geschikte instrumenten of methoden om kennisoverdracht van de opgedane ervaringen plaats te laten vinden. Bovendien gebruiken de meeste SAT locaties in ontwikkelde landen secundair en tertiair afvalwater, in tegenstelling tot ontwikkelingslanden waar het lastig is goed voorgezuiverd afvalwater te verkrijgen als gevolg van de hoge investeringen en operationele kosten. SAT met primair effluent (PE) is een aantrekkelijke optie voor ontwikkelingslanden omdat de behandeling van afvalwater tot op dit niveau kosteneffectief is en niet veel expertise vraagt van de beheerder van de RWZI. Toch is er weinig informatie beschikbaar over het gebruik van dit type afvalwatereffluent voor SAT. Daarom is onderzoek nodig om het lot van zwevende stoffen, organisch materiaal, nutriënten, organische microverontreinigingen en pathogenen onder verschillende omstandigheden tijdens de SAT van PE te begrijpen. Daarnaast is het van principieel belang om een raamwerk te vormen en ondersteuning te bieden tijdens besluitvorming, waardoor de uitvoering van de nieuwe SAT projecten succesvol kan worden ondernomen.

Laboratoriumschaal bodemkolommen en batchreactor-experimenten zijn uitgevoerd alsmede de analyse van de verzamelde gegevens uit de literatuur over laboratoriumexperimenten, pilot en full-scale SAT locaties om een duidelijk begrip van SAT prestaties vast te stellen. De effecten werden onderzocht van temperatuurverandering, redox, grondsoort, hydraulische belasting, voorbehandeling van PE, biologische activiteit en van natte en droge cycli op de verwijdering van geselecteerde verontreinigingen uit PE.

Laboratoriumschaal bodemkolommen werden gebruikt om het effect van PE voorbehandeling te bepalen (vóór de infiltratie van PE in de bodem) op de verwijdering van zwevende stoffen, bulk organisch materiaal (gemeten als opgeloste organische koolstof), nutriënten (stikstof en fosfor) en pathogenenindicatoren. Twee coagulanten, namelijk aluminiumsulfaat en ijzerchloride, werden getest. Experimentele resultaten toonden geen verschil in de totale verwijdering van zwevende stoffen bij infiltratie van gecoaguleerd en niet-gecoaguleerd PE (een stabiel verwijderingspercentage werd bereikt van ~ 90%). Echter, coagulatie-infiltratie verhoogde de verwijdering van bulk organische stof, fosfor en pathogenenindicatoren respectievelijk van 16 tot ~ 70%, 80 tot >98% en 2.6 tot >4 log10 eenheden (waarbij de lagere verwijderingswaarden werden bereikt tijdens alleen infiltratie). Beide coagulanten kunnen eveneens worden toegepast om de algehele prestaties van het systeem SAT te verbeteren en het benodigde grondgebied te verkleinen.

Het effect van de grondsoort en de hydraulische belasting op de verwijdering van bulk organische stof en stikstof werd onderzocht met behulp van een 5 meter lange bodemkolom, gepakt met kwartszand en duinzand. Geen significant verschil kon worden waargenomen in de verwijdering van opgeloste organische koolstof (~ 46%) wanneer de hydraulische belasting werd teruggebracht tot 1.25-0.625 m/dag. Echter, verwijdering van ammonium-stikstof was 50% hoger bij een hydraulische belasting van 0.625 mg/dag ten opzichte van een belasting van 0.625 mg/L. Bovendien was de ammonium-stikstofverwijdering in de duinzandkolom 10% hoger dan de verwijdering in de silicazandkolom. Geconcludeerd kan worden dat het SAT-systeem met relatief fijne bodemdeeltjes, die gestuurd wordt op een relatief lage hydraulische belasting, leidt tot een betere verwijdering van ammonium-stikstof. Niettemin vergt een dergelijk systeem veel onderhoud in de vorm van drogen en schrapen van het bodemoppervlak.

Verwijdering van zwevende stof, bulk organische stof, stikstof en pathogenenindicatoren uit PE werd onderzocht bij continue bevochtiging en wisselende periodes van bevochtiging en drogen met behulp van een 4.2 m lange bodemkolom. Geen significante verhoging van de verwijdering van zwevende stof (~

90%) en opgelost organische koolstof (50-60%) kon worden waargenomen bij een toenemende droogtijd. Toch werd een opmerkelijke stijging waargenomen in de verwijdering van ammonium-stikstof en pathogenenindicatoren bij toename van de droogtijd. Ammonium-stikstofverwijdering was gestegen van 20% bij continue bevochtiging tot maar liefst 98% bij een droogtijd van 6.4 dagen, terwijl verwijdering van E. coli en totaal-coliformen steeg van 2.5 log10 eenheden onder continue bevochtiging naar >4 log10 eenheden bij 6.4 dagen droogperiode. Samengevat kan gezegd worden dat de verwijdering van zwevende stof en opgelost organisch koolstof onafhankelijk was van de duur van de droogtijd, en dat de verwijdering van stikstof, E. Coli en totale coliformen verhoogd kon worden naarmate de duur van de droogtijd toenam.

De invloed van biologische activiteit op de verwijdering van bulk organische stof, stikstof en geselecteerde farmaceutisch actieve stoffen (PhACs) uit PE werd bestudeerd in laboratorium batchreactoren. Biologische activiteit (gemeten als adenosine trifosfaat, ATP) correleerde positief met de verwijdering van opgelost organische koolstof, die van 14% in blanco reactoren geleidelijk toenam tot 75% in de reactor met de hoogste biologische activiteit. Ook de verwijdering van ammonium-stikstof nam toe als gevolg van biologische activiteit, met 10-95%. Terwijl de verwijdering van neutrale hydrofiele verbindingen (octanol-water verdelingscoëfficiënt log Kow<2) van fenacetine, paracetamol en cafeïne onafhankelijk was van de mate van biologische activiteit, was >90% verwijdering van pentoxifylline afhankelijk van de biologische activiteit en de lengte van de reactorrijpingsperiode. Anderzijds steeg de verwijdering van gemfibrozil, bezafibraat en diclofenac van minder dan 10% in de blanco en controlereactoren naar >80% in biologisch actieve reactoren, wat een afhankelijkheid impliceert van biologische activiteit. Verwijdering van clofibrinezuur en carbamazepine van <50% in de meeste reactoren suggereert dat verwijdering van deze verbindingen niet afhankelijk is van biologische activiteit. Concluderend kan gezegd worden dat de verwijdering van opgelost organische koolstof positief samenhangt met de mate van biologische activiteit. Op dezelfde wijze nam de verwijdering van PhACs gemfibrozil, diclofenac, bezafibraat, ibuprofen, naproxen en ketoprofen toe met de mate van biologische activiteit, terwijl carbamazepine en clofibrinezuur continu werden gevonden, ongeacht de omvang van de biologische activiteit in de reactor.

Het effect van temperatuur en redoxpotentiaal op de verwijdering van bulk organische stof, stikstof, fosfor en pathogenenindicatoren werd onderzocht met behulp van laboratoriumschaal bodemkolommen en batchreactoren. Terwijl in de grondkolommen een gemiddelde opgelost organische koolstofverwijdering van 17% werd bereikt bij 5°C, verhoogde de verwijdering met 10% bij elke 5°C temperatuurstijging over het bereik van 15-25°C; bij 25°C werd een verwijdering van opgelost organisch koolstof bereikt van 69%. Bovendien vertoonden de aërobe bodemkolommen een verwijdering die 15% hoger was dan in zuurstofloze kolommen, terwijl de aërobe batchreactoren weer een 8% lagere verwijderingspercentages lieten zien dan de overeenkomstige anoxische batchexperimenten. Ammonium-stikstofverwijdering van >99% werd waargenomen bij 20°C en 25°C, terwijl de verwijdering aanzienlijk daalde bij 5°C, tot 9%. Hoewel ammonium-stikstof werd verwijderd tot 99% middels aërobe batchreactoren bij kamertemperatuur, resulteerden anoxische experimenten onder soortgelijke omstandigheden in 12% ammonium-stikstofverwijdering. In het licht van deze bevindingen zal het SAT systeem bij hoge temperaturen in de zomer een betere verwijdering laten zien van opgelost organisch koolstofverwijdering, stikstof, E.Coli en totale coliformen uit PE dan in de wintermaanden. Onvoldoende beluchting van het SAT-systeem als gevolg van een korte droogperiode kan leiden tot slechte reductie van ammonium-stikstof.

Huidige SAT sites die op dit moment over de hele wereld in werking zijn, hebben de neiging zich te richten op de operationele aspecten om aan de kwaliteitseisen van hergebruik te voldoen. Als gevolg hiervan is er te weinig aandacht besteed aan de ontwikkeling van evaluatie-instrumenten die op basis van de opgedane ervaring het implementeren van SAT technologie op nieuwe locaties makkelijker zou kunnen maken. In deze studie zijn zowel een raamwerk als instrumenten voor SAT implementatie ontwikkeld voor verschillende gebruikers, variërend van beleidsmakers en planologen tot ingenieurs en SAT-exploitanten. Een SAT voorbeoordelingsmethode behandelt de institutionele, juridische, sociaal- politieke en technische vereisten, waarna er voor identificatie, ontwerp, exploitatie en onderhoud andere methodes zijn ontwikkeld. Bovendien werd een waterkwaliteitsvoorspellingsmodel ontwikkeld om een schatting te kunnen maken van de mogelijke verwijdering van opgelost organische koolstof, stikstof, fosfor, bacteriën en virussen op basis van kenmerken van het afvalwatereffluent, het type voorbehandeling en de te overbruggen afstand. Het model is vooral handig om de behoefte aan nabehandeling te beoordelen op de waterkwaliteitseisen voor hergebruik, en helpt te voldoen in de raming van de totale investeringskosten die nodig zijn om eventuele nabehandeling in overweging te nemen.

Dit proefschrift onderzocht de mogelijkheden van het gebruik van de SAT-technologie voor de verdere behandeling en hergebruik van PE door middel van experimenteel werk en de ontwikkeling van evaluatie-instrumenten die geschikt zijn voor verschillende stadia van SAT, in combinatie met een waterkwaliteitsvoorspellingsmodel. Hoewel de methodes en het waterkwaliteitsvoorspellingsmodel werden ontwikkeld, getest en gevalideerd met behulp van gegevens uit laboratoriumexperimenten, pilot en SAT locaties gelegen in ontwikkelingslanden, zijn deze methoden en het model generiek en kunnen gemakkelijk worden aangepast aan andere locaties in ontwikkelingslanden. Het proefschrift biedt een uitgebreide methodologie die nuttig zal zijn voor beleidsmakers, planners en ingenieurs die nieuwe SAT locaties willen ontwikkelen en exploiteren, met name in ontwikkelingslanden waar SAT (met PE) is nog niet tot de volle potentie wordt benut.

LIST OF PUBLICATIONS

Articles published in refereed journals

Abel, C. D. T., Vortisch, R. C., Ntelya, J. P., Sharma, S. K. and Kennedy, M. D. (2014). Effect of primary effluent coagulation on performance of laboratory-scale managed aquifer recharge system. *Desalination and Water Treatment, In Press.*

Abel, C. D. T., Sharma, S. K., Mersha, S. A. and Kennedy, M. D. (2014). Influence of intermittent infiltration of primary effluent on removal of suspended solids, bulk organic matter, nitrogen and pathogens indicators in a simulated managed aquifer recharge system. *Ecological Engineering,* **64**, 100-107.

Abel, C. D. T., Sharma, S. K., Bucpapaj, E. and Kennedy, M. D. (2013a). Impact of hydraulic loading rate and media type on removal of bulk organic matter and nitrogen from primary effluent in a laboratory-scale soil aquifer treatment. *Water Science and Technology,* **68**(1), 217-226.

Abel, C. D. T., Sharma, S. K., Maeng, S. K., Magic-Knezev, A., Kennedy, M. D. and Amy, G. L. (2013b). Fate of bulk organic matter, nitrogen, and pharmaceutically active compounds in batch experiments simulating soil aquifer treatment (SAT) using primary effluent. *Water, Air, and Soil Pollution,* **224**(7), 1-12.

Abel, C. D. T., Sharma, S. K., Malolo, Y. N., Maeng, S. K., Kennedy, M. D. and Amy, G. L. (2012). Attenuation of bulk organic matter, nutrients (N and P) and pathogen indicators during soil passage: Effect of temperature and redox conditions in simulated soil aquifer treatment (SAT). *Water, Air, and Soil Pollution,* **223**, 5205-5220.

Journal articles submitted/in preparation

Abel, C. D. T., Al Kubati, K. M. A., Sharma, S. K. and Kennedy, M. D. (2014). Framework for site selection, design, operation and maintenance of soil aquifer treatment (SAT) system. *Journal of Water Resources Management,* Submitted.

Abel, C. D. T., Sharma, S. K. and Kennedy, M. D. (2014). Managed aquifer recharge for treatment and reuse of primary effluent: An overview. *Science of the Total Environment,* (In preparation).

Articles in conference proceedings

Abel, C. D. T., Sharma, S. K., Ntelya, J. P., and Kennedy, M. D. (2013a). Effect of primary effluent coagulation on removal of suspended solids, Bulk Organic Matter and Nitrogen during Soil Aquifer Treatment: Column Study. In proceedings of IWA Reuse Conference, October 27 - 31, Windhoek, Namibia.

Abel, C. D. T., Sharma, S. K., and Kennedy, M. D. (2013b). Removal of bulk organic matter, nitrogen and pharmaceutically active compounds from primary effluent in simulated soil aquifer treatment. In proceedings of IWA Reuse Conference, October 27-31, Windhoek, Namibia.

Abel, C. D. T., Malolo, Y. N., Sharma, S. K. and Kennedy, M. D. (2012). Effect of temperature and redox conditions on attenuation of bulk organic matter and nutrients in simulated SAT studies. In proceedings of IWA World Water Congress, September 16-22, Busan, South Korea.

ABOUT THE AUTHOR

Chol Deng Thon Abel was born in Malakal, South Sudan in 1976. He graduated from the University of Khartoum, Sudan in 2002 and holds a Bachelor of Science (B.Sc.) degree in Civil Engineering. From September 2002 to September 2005, Chol worked as a project Engineer/manager for various construction projects including irrigation pumping stations, open channels, water treatment plants and multi-storey buildings at Irrigation Works Corporation. In October 2005, he joined Lahmeyer International – Consulting Engineers, Germany as soil laboratory engineer during the construction of Merowe dam (1250 MW). From April 2007 to February 2010, Chol assumed the position of Assistant Coordinator for South Sudan hydropower projects as part of Dams Implementation Unit (DIU) of the presidency of Sudan where he participated in preparation of terms of reference, tender documents and evaluation of contracts for the new hydropower projects.

In October 2007, Chol was awarded a scholarship from the Joint Japan/World Bank Graduate Scholarship Programme (jj/wbgsp) to study at UNESCO-IHE where he obtained a Master of Science (M.Sc.) degree in Municipal Water and Infrastructure (specialization in Water Supply Engineering) in May 2009. In March 2010, he started his full-time Ph.D. research entitled: "Soil Aquifer Treatment: Assessment and Applicability of Primary Effluent Reuse in Developing Countries" at UNESCO-IHE and Delft University of Technology under the financial support of UNESCO-IHE Partnership Research Fund (UPaRF) through NATSYS project.

During his Ph.D. research, Chol followed the educational programme of the Netherlands research school for Socio-Economic and Natural Sciences of the Environment (SENSE) where he fulfilled all requirements and obtained a certificate in June 2014.

THE SENSE RESEARCH SCHOOL CERTIFICATE

Netherlands Research School for the
Socio-Economic and Natural Sciences of the Environment

C E R T I F I C A T E

The Netherlands Research School for the
Socio-Economic and Natural Sciences of the Environment
(SENSE), declares that

Chol Deng Thon Abel

born on 23 February 1976 in Malakal, South Sudan

has successfully fulfilled all requirements of the
Educational Programme of SENSE.

Delft, 17 June 2014

the Chairman of the SENSE board   the SENSE Director of Education

Prof.dr.ir. Huub Rijnaarts   Dr. Ad van Dommelen

The SENSE Research School has been accredited by the Royal Netherlands Academy of Arts and Sciences (KNAW)

K O N I N K L I J K E   N E D E R L A N D S E
A K A D E M I E   V A N   W E T E N S C H A P P E N

Printed and bound by CPI Group (UK) Ltd, Croydon, CR0 4YY

21/10/2024

01777096-0009